The Body Quantum

THE
Body Quantum

◈

The New Physics of Body, Mind, and Health

FRED ALAN WOLF

MACMILLAN PUBLISHING COMPANY
New York

MACMILLAN PUBLISHING COMPANY
866 Third Avenue, New York, N.Y. 10022
Collier Macmillan Canada, Inc.

Library of Congress Cataloging-in-Publication Data
Wolf, Fred Alan.
The body quantum.
Bibliography: p.
1. Quantum biochemistry—Philosophy.   2. Biophysics.
3. Health.   I. Title.
QP517.Q34W64   1986   610   86-8403
ISBN 0-02-630890-8

10 9 8 7 6 5 4 3 2 1

PRINTED IN THE UNITED STATES OF AMERICA

CONTENTS

FOREWORD

Something peculiar is happening in modern medicine. Everywhere today there are events that seem important but that are being ignored—happenings that cry out for an explanation, yet from which we hide. Why, for example, are we not captivated by the astonishing fact that more heart attacks, the commonest cause of death in our society, occur on Monday than any other day? Why should man be the only mammal who dies more frequently on a given day of the week? Further puzzling events are not hard to find. For instance, why do patients recovering from gall bladder surgery have shorter hospital stays if the windows in their rooms open to a view of trees instead of a brick wall? Why does the immune function in widowed spouses stop working, and why does their death rate escalate in the first year of bereavement? Why do patients with angina have 50 percent less pain if they perceive their wives to be loving and caring? Why do the *majority* of patients who have their first heart attack in this country display *none* of the major risk factors for heart disease—cigarette smoking, high blood pressure, diabetes, or elevated cholesterol levels in the blood?

Perhaps it is not quite right to say that these anomalous events are ignored, for they have all been documented by careful scientific observation, and they are indeed attracting the scrutiny of an increasing number of interested investigators. But as yet, these types of events, which seem persistently tied to the fact of human consciousness, are of little interest to most bioscientists and practicing physicians. The focus of medicine remains to this day the physical body, and clinical events which smack of "the mind" are somehow considered not as legitimate or as "real" as physically based problems.

In *The Body Quantum*, physicist Fred Alan Wolf introduces a stunning new model into medicine that may allow us to re-envision the body, as well as help us solve such clinical puzzles as those posed above. Drawing on the insights of modern physics, he develops a new view of the body and of health and illness. In this new scheme of things, it is impossible to even speak of "physical" illnesses as if they could stand apart from the mind and consciousness. As he puts it, the physical world of hard matter, light, and energy simply does not and cannot exist independently of human consciousness. Central to this novel view of the human body is the crucial role that the observer plays in quantum physics, without whom much of this modern picture of the world cannot be understood.

But why should we listen to Dr. Wolf? Why not continue with "business as usual," with our current ideas of how the body functions, which solely rely on classical views of matter and energy? Why introduce something as ethereal as consciousness to explain the function of something as concrete as the physical body? There are several reasons which go beyond the clinical observations above.

Consider the prediction of one of the greatest physicists of this century, Niels Bohr, who stated that the eventual addition of biological concepts to quantum physics was a "foregone conclusion." In the same vein, there stands the observation of the scientist-philosoper Jacob Bronowski that quantum-based events are just the kind of happenings that go on in our brains and our nerve tissue, and in the DNA that makes up our genetic material. Or the statement by the contemporary physicist John Archibald Wheeler, that at bottom the entire world is a quantum world, and that any system (including, we may presume, the human body) is ineradicably a quantum system.

Against these sorts of observations, which predict that someday medicine will have to come to terms with quantum physics, stand the self-confident stances of most persons in medicine who know little about the stunning implications of modern physics. The situation is, with all due respect to the incredible fruits of modern bioscience, schizophrenic: On the one hand, the most accurate science we have ever had—that of modern physics—includes the role of the observer in its most widely accepted interpretation, while modern bioscience excludes such a role by insisting that human bodies can be understood in classical terms. The situation is strained, to say the least. And it *cannot be resolved* by the usual appeal that bodies, as macroscopic objects, are oblivious to quantum-sized events, that these effects are "washed out" through the laws of large numbers, and that consciousness "therefore" has no role in bodily function.

Fred Wolf perceives this dissonance and has provided an epochal book that is a major attempt to introduce quantum thinking into medicine in a comprehensive way. It is important, too, that it is a physicist such as Dr. Wolf, who has strong academic credentials, who has performed this service; for these insights are from one who lives and breathes the rarefied air of the mathematical heights at which quantum physics soars.

To those who view Wolf's attempt to reshape our understanding of the body as a trivial excursion into unnecessary complications, it should be said that nothing is farther from the truth. Our models of reality determine in very real ways what we *can* observe, what we perceive to be important, and what we decide to do about certain problems. Nowhere is it more important to realize this than in the world of medicine, where human suffering and human lives are at stake. We are learning that what we cannot conceptualize we cannot implement—and so our hands remain tied, therapeutically speaking, to the time-honored pillars of drugs and surgery, and we are prevented by our mind*less* models from executing valuable therapies that rest on consciousness.

Dr. Wolf's book is a major contribution to medicine; it is a major call to a new vision; above all, it is an attempt to reintroduce human consciousness into the body at a time when reductionistic science has done its best to take it out. And it is a signal that we confront a new beginning in medical thought, one that is as radical as when science first entered medicine as an unexpected guest. It is a portent of things to come—of *mind*-things, of *consciousness*-things, which, of course, are really not things at all.

Please be reminded, dear Reader, that as you venture into *The Body Quantum* you are not venturing into some *Gray's Anatomy*: This is stronger medicine. It cannot be digested with the same world view or outlook that proves so useful in the anatomy or physiology laboratory. *The Body Quantum* is about something more than the body—and that "something more" is the message of the book.

—Larry Dossey, M.D.
Dallas, Texas
June 1986

PREFACE

Whenever I have read a book or paper by Fred Alan Wolf, I sense that all my neurons have become activated and are firing at maximum. Although not expected at my age, I feel that new connections—synapses—are being formed and that I am entering a new and higher level of comprehension of how a human being functions, and the nature of the "physical" world. It is not that I agree with everything he writes or that I fully understand every equation, but because of his felicitous style and rare ability to make complex physics comprehensible, I can follow most of what he has written. My immediate thought is, "This is what I have been looking for. This is a new way of thinking about complex problems whose resolution appeared insoluble."

I am reminded of a statement by Albert Einstein, in a somewhat different context, that "the splitting of the atom has changed everything except our way of thinking, and thus we drift toward unparalleled catastrophe. . . . We shall require substantially a new manner of thinking if mankind is to survive." The phrase "a new manner of thinking" comes to mind in reading this book with regard to the future development of psychiatry, the behavioral sciences, and medicine. For some time medical doctors and psychiatrists, like myself and the noted Dr. George Engel, have looked for such a new biomedical model with the help of ideas from modern physics.

In *The Body Quantum*, Fred Wolf utilizes the principles of modern physics to explain and interpret the workings of the human body in health and disease. Included among the many topics that he covers are the mind-body interaction and a new and challenging view of consciousness in which a profound connection is made between human consciousness and quantum physics. In his explanation of consciousness—the sensation of thought, emotion, and awareness—Wolf utilizes the most generally held view of quantum mechanics, the so-called Copenhagen interpretation that maintains that the act of observation, the intrusion of human conscious activity, is required before an actual recognizable event can occur at the subatomic level. The observer effect means that the actions of a human observer play a more "dynamic role in the universe than was previously suspected."

Fred Wolf points out a well-known fact in physics that "just when a particle is a solid object and when it is a wave appears to depend on that certain unpredictable action known as the observer effect." Thus, if observations are carried out in one manner, a subatomic particle may appear as a solid object. But in another kind of observation it may appear as a physical wave. It is not only that the act of observation or the manner in which the observation is made alters the nature of that which is observed, but it is impossible to determine the position and momentum of such a subatomic particle at the same time, an illustration of Heisenberg's principle of uncertainty. This notion of "complementarity" is only one of many encounters one meets in describing the behavior of the world that is in violation of our everyday experience as well as the teaching and practice of medicine.

By far and large these are notions that have been accepted in physics for over eighty years. Yet, paradoxically, they have not been incorporated in any of the thinking of medical scientists, physiologists, biochemists, and others who rely on Newtonian mechanistic, linear causal concepts. To them, the mind is located in Euclidian space. But for Wolf, consciousness is identified with the process of wave transformation through setting tolerances for observing either energies or locations of protein gate molecules imbedded in the neural membrane.

However, my colleagues might very well ask, Why bother with all this complicated material? We're doing very well as we are. Progress in medicine and particularly biological psychiatry is most impressive.

Although we pay lip service to the concept of wholeness and the importance of multiple variables, we have no model—no way of expressing the complexities—of the human being and especially of the mind-body relationship and consciousness. It is readily acknowledged by most (although not all) that a biologic reductionist model, or for that matter a psychologic or social reductionism, is inadequate. But having said that, we end up by saying that for heuristic purposes we must confine our studies to one or another variable. We find it impossible to put all the elements together into a cohesive whole.

Fred Wolf's work with quantum mechanics makes another aspect more evident: The more one probes, the more one penetrates to smaller and smaller particles. This phenomenon can be seen in current medical and psychiatric research. The target to which the research is addressed has moved from the organ, to the cell, to the molecule, to the atom, and now to the subatomic particle. In this process, does one increase reliability or accuracy? Do we necessarily refine and make more precise our knowledge and gain more control?

When one reaches the subatomic world, however, we encounter the principle of uncertainty which sets limits on what we can know. It is not a matter of inventing a new technique to overcome this principle; it is a built-in quality of the world that we can never exorcise. But so many of our medical and psychiatric researchers are working ahead, oblivious of this obstacle before them.

At the other extreme, in order to comprehend the universe and other universes, modern physics is necessary. Similarly, to conceptualize the wholeness of a human being, modern physics might also be necessary. As Wolf puts forth, modern physics might give us a new manner of thinking that can lead to more success in our endeavors and aid us in providing a comprehensive view of the workings of man.

Dr. Larry Dossey has pointed out that the constraints of the classical approach have led to dehumanizing notions of what human beings are all about, views that are hard to defend from the perspective of modern physics. He quotes Carl Sagan, who observed that the "fundamental premise about the brain is that its workings—what we sometimes call 'mind'—are consequences of its anatomy and physiology and nothing more." To continue Dossey's comment, "As Wolf's perspective illustrates, there are points of view that flow from modern physics in which this 'fundamental premise' may not be as fundamental as it seemed, and the 'nothing more' may hold some surprises."

The implication is clear: Classical mechanistic fragmented approaches necessarily end up in fragmenting the human and thus dehumanizing the individual. We see this in medicine in preoccupations with organs: "This man has an interesting liver with cancer." Nothing about the person—his history, his work, his family, his aspirations.

Thus, we see that quantum mechanics and concepts of modern physics with relativistic, probabilistic, and nonlinear causal views have application not only for electrons, other subatomic particles, and for stars and galaxies, but also may be of critical importance in the workings of the body in health and disease. In this way, the work of Fred Alan Wolf may open up a new manner of thinking and consequently achieve a level of physical and psychological health hitherto not attainable.

—Alfred M. Freedman, M.D.
Professor and Chairman, Department of Psychiatry
New York Medical College, Valhalla, New York
Editor-in-Chief
Integrative Psychiatry
June 1986

ACKNOWLEDGMENTS

Not being a physician, biologist, neurophysiologist, or anatomist did pose challenges in attempting a book such as this one. I tried as a physicist to apply skills well honed in one area of science to illuminate the continuing mystery of the body. For any errors, omissions, and plain misunderstandings regarding this magnificent body human we all inhabit, I take full responsibility. However, I did ask friends and colleagues to lend a hand by reading the manuscript and making corrections where they sorely were needed.

To John W. Travis, M.D., M.P.H., I am extremely grateful for many helpful comments, corrections and knowledge.

To Larry Dossey, M.D., I appreciate his enthusiastic support and comments after reading a first draft, and for writing the Foreword.

To Alfred M. Freedman, M.D., thanks for writing the Preface.

To my editor, Charles Levine, and copyeditor, Tony Davis, at Macmillan, I am grateful.

But most of all, I owe a great deal to Catherine Shaw, who painstakingly, and I mean painstakingly, line edited the manuscript, not once but twice full over. She dug into the "guts" of the book and helped me reform it with many suggestions, insights, and profound questions ("queries" in the lingo of the publishing business).

I also want to thank Michael Talbot and Mr. and Mrs. Abraham Levine for their hospitality when I was visiting New York to complete the manuscript.

To all of my friends my deepest gratitude.

Fred Alan Wolf
New York
July 1986

Introduction: The Body Is a Mystery

If you take any time at all to think about your body, you probably are just as mystified as I am that it exists. With the millions upon millions of processes occurring inside each of us, it appears overwhelming that the body works the way it does. I mean, why does life express itself as a human body that is moving, eating, building, sensing, breathing, minding, healing, and transforming itself?

The bodies we all inhabit are capable of vast spectrums of experience. This marvelous instrument can become elated, depressed, exhausted, or exalted. It can express itself through enjoying, growing, thinking, feeling, and loving. It can also feel hate and anxiety, become overworked and overstressed, and feel sexual desire. It can heal itself or kill itself. And seemingly unlike other living creatures and plants, we know when we are experiencing life. And we know that life in our bodies must eventually come to an end. Death still holds the ultimate mystery.

Probably even the earliest humans were aware of these mysteries. Did they wonder why they were alive rather than simply dead objects floating in space and moving eternally from the beginning of time to the supposed end of all existence? Perhaps they did.

If you are like me, your mind is forever exploring, looking under the rocks of your own timeless existence and asking, Why am I here? Why are there living things rather than just dead objects floating in space? How are life and the lifeless related? How can we explain consciousness?

Ancient Physicians and Physicists

Such questions were certainly posed by the ancient Greeks. Aristotle, for example, believed that the dividing line between life and the lifeless was invisible. He wrote:

> Nature proceeds little by little from things lifeless to animal life in such a way that it is impossible to determine the exact line of demarcation, nor on which side thereof an intermediate form should lie. (McKeon, p. 249.)

In the days before the 17th century, at the dawn of the Age of Reason, the physicist and the physician were the same person. Matter was not necessarily viewed, as it is by many of today's physicists, as dead inert stuff animated by forces and nothing more. Somehow, as Aristotle had put it, one still did not know "on which side thereof an intermediate form [of life] should lie." But in the 17th century, René Descartes pronounced the division of body and mind. The body was seen as a separate entity from "something else" called the soul. Thus, the soul was believed to be the entity that gave life to the lifeless body composed of inert stuff.

Somehow the soul "animated" the dead matter of the body, moving it mechanically through space and along the inevitable course of time. From Isaac Newton came the idea of mechanical time, limitless space, and dead inert matter. And by the end of the "flash" of the 100 years of history called the Age of Reason, the whole universe was seen as a machine.

So was the human body. The physician, now specialized and separated from the physicist, peered into the body, looking for clockwork in its workings and rhythms. Much was revealed, yet certain mysteries remained.

Modern Medicine and Physics

Today, although we have accumulated a great amount of medical and biological knowledge, the body human still holds its mystery. It seems that the more we learn the more we have yet to learn.

Many books about the body are presented as a kind of tour. One can almost hear the tour guide barking, "Look over here, ladies and gentlemen. Here you see the liver and the marvelous workings of its machinery. And over here, step lively please, we don't have much time, is the nervous system, and over there is . . ." Such a tour of the body may be useful, but it hardly gets to the meat of the matter: the twin mystery of how life both takes root in seemingly dead, inert matter and also somehow "knows" that it is doing so. These are the questions I intend to explore in this book.

Even though modern science has provided a great number of detailed maps of the body, the experience we call life is no clearer now than it was centuries ago. However, I have lately begun to wonder if the age-old dream of uniting the soul of life and the body of inert matter is possible through the new physics. By the new physics I mean the physics discovered during our century, specifically, quantum physics. Despite its amazing insights, most research concerning the processes of life fails to take quantum physics into account.

Quantum physics deals with the movements and changes of matter and energy at the tiniest levels of existence we know of: the world of atomic and subatomic matter. Probably the most important discovery of quantum physics was the *observer effect*. This showed that how an observation was performed—what the physicist brought to bear on the observation of matter—disturbed that matter in uncontrollable ways. This disturbance wasn't simply the result of an error caused by a maladroit experimenter. It involved, instead, a fundamental new discovery that there *had* to be unexpected results, no matter how carefully the physicist performed the experiment. This was due to a new principle at work in the physical universe—the Heisenberg uncertainty principle.

This principle showed that our ordinary vision of the material world—as composed of particles of matter following laws of motion—was fundamentally in error. Instead of this, atomic and subatomic matter violated our ordinary sense perceptions. For example, a subatomic particle could not have a well-defined location in space and also follow a well-defined path in time. One or the other had to go. A well-defined position for a subatomic particle meant that it had no clear path into the future. Conversely, a clear motion of the particle toward the future meant it had no location in space.

To resolve this paradox, a human experimenter was needed. The experimenter had to choose what to observe, either a particle's location in space or its path through time. In this manner, the experimenter was "creating" the reality of the subatomic or atomic particle by his or her choice. This meant that the actions of a human observer played a more dynamic role in the universe than was formerly supposed.

The old physics of Galileo and Newton provided clues to how any object can move and relate with other objects. It made the mechanics of the body possible to understand. But it didn't solve the mystery of life. Nor could it explain how conscious life can occur.

Certainly, I thought, the discovery of the quantum nature of all matter—how the observing of matter fundamentally alters its course—must shed light on how the body can be alive in the first place. So far, medical books, body books, health books, and other books related to health and illness appear to treat the body as a machine without looking into the wider question, Why is this machine alive and thinking? Could the discoveries of the new physics provide any insight into this mystery?

Furthermore, with all the information available in both medical publications and popular books about the body, surprisingly little has been written about the physics, either old or new, of the body. Yet all bodily processes must be governed by the laws of physics. Healing, digestion, sexual function, and even daydreaming must, somehow, be inextricably caught up and governed by physical laws.

Thus, physics, both classical and quantum, must enter the body human. The division between living and dead inert matter must be reexplored. It is hoped that the discoveries of quantum physics, specifically, the effect of human observation on all matter, will play a role, perhaps a role far more significant than has previously been suspected. To this end this book is written.

Physics in the Body

Our bodies move on temporal and spatial scales both beyond and within our control. The automatic processes—those that occur without our conscious control—are probably the most mysterious. Our organs function in a complex dance of physical movement coupled with electromagnetic impulses. Our bodies respond and produce forces resulting in balance and motion. Our bones are designed to support just the right amount of weight, within well-designed engineering tolerances—no more, no less.

Without physical forces our bodies wouldn't move, our lungs breathe, or our hearts pump. As it is, our blood supply moves at just the right speed with little turbulence: Any faster and the greater turbulence would result in heart attacks and strokes; any slower and we'd die of oxygen starvation.

We walk, bend, dance, lift, and throw the way we do because of Newtonian forces. We breathe air following basic laws of gases, such as the adiabatic law of pressure. Our bodies are warmed and cooled according to the Stefan-Boltzmann law of heat radiation. Weight loss and gain are determined by the law of conservation of energy. That physics underlies all molecular processes in the body appears unquestionable.

In this book we shall examine the body using the eyes of both classical and quantum physics. Although classical physics was long ago replaced by the underpinnings of the new physics called quantum physics, classical physics is still very useful in explaining both the mechanical and the electromagnetic processes taking place both inside and outside the body. For example, it is not necessary to determine the quantum physics of bones in order to understand how your skeleton holds you up or why it breaks under stress. Nor is it necessary to use quantum physics to understand how a nerve conducts electricity. These and many other processes can be understood quite well in terms of classical physics.

Introduction

Classical and Quantum Physics

The fundamental flaw of classical physics is that it is deterministic. It implies that from a given cause a specific effect can be predicted. We have gained from the new quantum physics the insight that this cause-effect relationship no longer holds at the level of atoms and molecules. We cannot predict an effect, such as when an energetically excited atom will radiate away that energy, or when a molecule will undergo a molecular change when it is hit by a particle of light. Even our picture of atoms as solid little things has changed—they are really more like clouds. The shape and form of these clouds depends on how we perform experiments on atoms.

When large numbers of atoms congregate, they form molecules. The larger the molecule, the more predictable it is, but only so long as we look at the molecule as a whole entity without attempting to look at its internal atomic structure.

Thus, classical physics cannot explain how a molecule of DNA is constructed or why certain kinds of diseases such as cancer or emphysema can arise. Nor can classical physics explain how we think or how our thoughts affect our bodies. I believe that quantum physics can shed light on these deeper mysteries.

Why We Need Quantum Physics to Explain the Living Body

Classical physics does not entirely explain the body; we also need quantum physics. Ultimately, the hidden mechanisms of the body are quantum mechanical and involve the cell's molecular biology—that is, the structures of DNA, proteins, fats, and carbohydrates, and the transportation of oxygen and carbon dioxide needed for the body's energy production. Possibly even the nature of the chronic universal disease—aging itself—can be understood, and cured, with the insights of quantum physics.

Through the eyes of quantum physics, one can see that the mind also begins to emerge as evidence of the ancient "soul"—that which governs and regulates the invisible atomic and molecular processes of life. These processes govern the living movement of matter in the body, both consciously and unconsciously. Here I put forward that the mechanism for governing the living movement of bodily matter arises through the effects of quantum physics, specifically, the effect that observation has on matter.

The effect of observing one's own body, both consciously and unconsciously, alters the body. Here, the words "unconscious observation" may appear mysterious. How can anyone observe anything unconsciously? I use the word "observation" in perhaps a different sense than to which you may be accustomed. In quantum physics, observation means any action that causes a choice to be made, for example, between the location of a particle in space and its movement through time. Quantum physics has shown us that such actions are quite sudden and discontinuous, disrupting previous patterns of behavior. Unconscious observations occur at every moment in our bodies; they are necessary in order to deal with the countless changes our bodies are put through in the simplest actions, getting up and walking across the room, for instance. Although we perform such actions unconsciously, I propose that they nevertheless involve *observations* at the level of quantum physical processes, and that without them we would be automaton machines. Unconscious observation is automatic, or, in medical terms, *autonomic*. Such processes as growth, body size, nervous function, and disposition are governed by processes probably not within our immediate conscious control. Conscious observation, such as when you notice you are feeling anger or love, assuredly alters your body chemistry and probably the movement of nerve impulses. Whether you are relaxed or tense can be modified by conscious observation of muscle groups, such as those in your neck and back, perhaps through the use of biofeedback techniques. Just how this works falls within the realm of quantum physics, specifically, the observer effect that says *how* you observe changes *what* you observe.

A Crash Course in Quantum Physics

In order to grasp the significance of quantum physics in the body, and specifically the *observer effect*, we need to look at a few facts of life—quantum physical life. Quantum physics deals with the movements and energies of atomic and subatomic particles such as the atoms and electrons coursing through our bodies. These particles are not simply microscopic billiard balls bouncing around through the actions of atomic forces. They are, instead, "tendencies" to exist sometimes as billiard-ball objects and at other times more like waves. In fact, just when a particle is a solid object and when it is a wave appears to depend on that certain unpredictable action known as the observer effect.

Briefly, whenever a particle is unobserved, it moves as a wave spread over a region of space. This wave, however, is not a physical wave like an ocean wave or a sound wave; instead, it is a "wave of probability." Each part of the wave represents a possible location for the particle. I call such a wave a "qwiff," standing for a *quan-*

tum wave function.

Quantum physics provides us with a mathematical description of qwiffs. Physicists can predict just how a qwiff will behave. The mathematical laws governing their movements are the same as the mathematical laws governing certain types of wave movements. That is why physicists named them quantum wave functions.

However, I repeat, these waves are not physical. They cannot ever be observed as waves. Instead, whenever an observation is carried out, the wave is said to "collapse" into a tiny region of space and appear as a particle. Any single observation always produces a particle; a series of observations will produce different patterns. These patterns are then later interpreted as either wave or particle motion.

If the observations are carried out in one manner, the pattern appears as a disjointed movement of the particle from one location to another. In such a series of observations, one cannot predict in advance just where the next observation will occur. In another kind of observation, a new pattern appears, reflecting that movement as a physical wave. Here, one can predict the overall pattern, and one must interpret that prediction in terms of a wave motion.

For example, in a famous experiment, known as the double-slit experiment, a single particle moves from a source, such as a hot filament, to a screen, such as a photographic emulsion plate. There it lands and makes a pointlike spot. The experiment is repeated several times, resulting in a distribution of random spots on the plate. Even though the experiment is repeated as exactly as possible by the experimenter, no one can predict just where any particular spot will appear. However, the cloud of spots on the screen can be predicted to follow a pattern like a wave washing across the plate. This wave pattern, however, is not in itself convincing. One could explain the pattern just as well by a series of particles landing on the plate randomly.

Next, the experiment is changed by placing a screen with two long, narrow, parallel slits between the source filament and the plate. Now each particle must pass through one or the other slit in order to reach the plate. In each single particle emission, a spot either appears on the plate or it doesn't. After several spots are observed, a new pattern begins to appear on the plate. It is called an interference pattern. It is not simply the result of random spots appearing on the plate produced by a series of particles passing through one of the slits; instead, there are regions on the plate that remain free of spots. If one calculates where these regions occur, one finds that they correspond to regions that would be produced by a wave that spreads over the two slits before arriving at the plate.

However, the time interval between particle emissions from the filament can be controlled so that only one particle passes through the slits at any given time. Yet a wave interference pattern still appears. Somehow, each particle passes through both slits at the same time in order to not arrive at the blank regions on the plate. The particle thus is said to move through the slits as a wave.

This paradox is called the "wave-particle duality." Whether one interprets the result as a wave or a particle depends on how the experiment is carried out. With no double-slitted screen between the source and the plate, the series of spot observations can be interpreted as a series of particle observations. When the double-slitted screen is in place, the pattern of spots must be interpreted as being produced by waves. Yet each single observation of a spot is still a particle observation. With just one observation, that of a single spot, one cannot ever observe the wave effect; but with a series of observations, the wave-interference effect can be observed.

The best description of this is to think of an unobserved particle moving through space as a wave. Whenever it is observed, it's a particle. Its wavelike history, however, is manifested by just how the experimenter sets up the series of observations. Thus, every particle of matter exists as a wave tendency when not observed, and as a particle reality when observed.

This causes the actions of observation, particularly the experimenter's setup, to play a major role in quantum physics. This is the action of the observer effect. I attribute it to human consciousness in particular, and to an overall consciousness in general. This overall consciousness, the experiment setup we call life, is called different names by different organizations of thought. Some call it God, others label it Evolution.

Through such choices—wave or particle—different physical events or series of events will occur. In our own bodies, there are many of these choices, and this is where quantum physics enters the body human. And thus I come to the *body quantum*.

What Is the Body Quantum?

The word "quantum" refers to a whole amount of something. Thus, the body quantum refers to a whole amount of something important governing the whole human body. That something is consciousness. It is my contention in this book that consciousness acts in a quantum manner inside our bodies. Its acts produce all the various activities we enjoy that make up our lives, be they extraordinary or dull.

Thus, quantum consciousness is the observer effect in quantum physics. What this means is that the action of simply observing something alters the thing observed in a sudden and disruptive manner. This change is reflected in quantum physics by changing that which is probable into that which is certain.

In so doing, events which have only tendencies to exist become actual events. Such events are marked by the entering of the body quantum into physical form. Our everyday lives are made up of seemingly countless quantum consciousness acts, movements of the body quantum from a stream of consciousness into the physical stream of the body itself. Every action we take, moving, eating, building the body, sensing, breathing, minding or noticing the body, healing and transforming the body, is a result of the body quantum—something probable has radically changed into something actual and conscious, which has also changed. For the body quantum cannot just affect the physical. It also affects our minds and thoughts. As we become more and more aware of ourselves, we also must alter the bodies we become aware of. These alterations can produce both blessings and curses, as we shall see in this book.

The Mind-Body Interaction

Because we humans can think and are able to gain conscious control of our lives, the mind-body interaction is extremely important and necessary for our understanding of the body human. It is my contention that by understanding the physics of the body, specifically, how quantum physics and the observer effect are involved in the body human, we will be able to gain healthier and happier lives.

To bring this in focus, I have divided the book into eight parts: Moving, Eating, Building, Sensing, Breathing, Minding, Healing, and Transforming. In each part I hope to show how both classical physics and quantum physics play essential roles in our comprehension of our bodies. In each part we will see both the over-structure provided by the classical physical understanding and the understructure provided by the quantum physical understanding—specifically, how consciousness, in its quantum role as the observer, alters the body, enabling each bodily function to occur.

PART ONE

Moving

In the famous Beatles' song "A Day in the Life," about waking up, getting out of bed, dragging a comb across one's head, the singer reminds us of the drudgery of simply moving our bodies and our arms from one place to another, particularly in the morning when hardly awake. Indeed, moving when we are just waking up seems to require conscious effort, while, later in the day, we move from place to place without giving it much thought at all.

Most of us take for granted the movements we perform each day. Yet a little memory reminds us of our first steps as toddlers and their enormous challenge. Sometime between infancy and childhood, moving our bodies through space changed from a challenge requiring thought and concentration to a habit of unconscious actions requiring no thought at all. As we aged, we found that other motions became possible, particularly if we took up athletics or dance. I remember the first day I began to learn to throw a football: Just getting the ball to "spiral" as it left my hand was quite a challenge. At the age of eight, I first learned to ride a bicycle. It seemed impossible to me at the time: How could a two-wheeled vehicle maintain its upright position as it moved along the road?

As adults, we find most of the movements we perform take little or no thought. Most of our movements are seemingly unconscious. However, consciousness is at work. Our bodies move not only by using our arms and legs but also our mouths, eyes, tongues, fingers, toes, and, if we take

1

up any physically demanding activity, such as skiing or yoga, through new combinations of muscles acting in accordance with each other. Peering into the body, we find other seemingly unconscious movements. Our intestinal tracts move in rhythmic undulations that pass our food along. Our hearts and diaphragms move periodically, again without any conscious actions on our parts.

How do all of the motions our bodies take part in occur? In this first part I hope to provide some insight by grasping a few basic principles of motion—from the world of physics.

The many influences on our bodily movements are spelled out in the next six chapters. As we shall see, underlying all of the movements of our bodies there exists a form of consciousness. And that consciousness, even though it is "unconscious" for the most part, requires intelligence, and, if my speculations are correct, the effect from quantum physics known as the observer effect described earlier. Thus, I hope to elucidate both the classical and the quantum physics of simple everyday bodily motion. In fact, this will not only help us to understand how we move, in part and in whole, it will also provide us with a better understanding of the differences between classical physics and quantum physics.

CHAPTER ONE

Some Assorted Body Spaces and Their Problems

How an Animal's Size Determines Its Movements

No doubt we have all been scared silly by movies featuring huge, ferocious beasts—giant spiders, moths, or prehistoric monsters, each one more horrifying than the last. Perhaps you wondered, could any of those beasts really exist? Could an ape really grow to the size of King Kong? Could a gargantuan ant stalk the plains of New Mexico as it did in the movie *Them*? What about an amazing colossal man? Could a human being exist around 60 feet in height with a girth and weight to match?

To answer these questions, we need look no further than the laws of classical physics. It is perhaps no surprise that these laws determine how our bodies move. But have you thought about how they shape and set limits on the physical size of all things, particularly animals?

Interestingly enough, Galileo had already thought about the problem of the size of animals even before the related laws of physics were formulated by Sir Isaac Newton. Galileo wrote:

> From what has already been demonstrated, you can plainly see the impossibility of increasing the size of structures to vast dimensions either in art or in nature; . . . nor can nature produce trees of extraordinary size because the branches would break down under their own weight, so also would it be impossible to build up the bony structures of men, horses, or other animals so as to hold together and perform their normal functions if these animals were to be increased enormously in height; for this increase in height can be accomplished only by employing a material which is harder and stronger than usual, or by enlarging the size of bones, thus changing their shape until the form and appearance of the animals suggest a monstrosity. This is

3

perhaps what our wise poet had in mind, when he says, in describing a huge giant:

"Impossible it is to reckon his height
So beyond measure is his size."   (Haldane, p. 1.)

In 1928, J. B. S. Haldane, a British geneticist who was professor of biometry (which means the measure of biological systems) at University College, London, wrote a delightful book called *Possible Worlds*. In it, Haldane pointed out that for most animals there is an optimal size. A rabbit could not be as big as a hippopotamus, nor could a whale be as small as a herring. As the size of an animal changes, so must its form.

Take, for example, the comparison between a gazelle, a creature with long thin legs, and a rhinoceros. Could a gazelle increase its size proportionally until it was as wide as a rhino? The answer is no, both because of the demands of gravity and the limits of bone structure. Let's look closer.

In order to triple its girth and proportionally increase its other dimensions, the gazelle would need to increase its volume by a factor of 9. Since we don't want the animal to be any heavier, that is, denser than necessary, its weight would also increase according to its volume; thus, the 100-pound gazelle would need to weigh in at 900 pounds. But this much weight would require a differently shaped body. Such a blown-up creature would no longer be a gazelle.

Similarly, if we picture a giant like the one in "Jack and the Beanstalk," say around 60 feet tall, and we wish him to be proportionally the same as an ordinary human, then his corresponding dimensions of width and thickness must also increase (otherwise he would be much too skinny). Since his height has increased by a factor of 10 (6 feet to 60), his waistline would also need to increase from about 36 inches to 360 inches—again the same factor of 10. But this means his volume will need to increase by an enormous factor of 1,000! Instead of weighing about 200 pounds, he now must weigh 100 tons! This means that every square inch of his bones, particularly his ankles, will need to support 10 times the weight that the bones of a normal 6-foot-tall man supports. Consequently, the giant would break his ankles after taking one step.

The simple presence of gravity determines how big an animal can be. A giraffe is taller than a hippo or rhino, and certainly bigger than a gazelle. It compensates for its thin legs as much as possible by compressing its body and stretching out its legs at oblique angles to the ground. These physical accommodations both increase the giraffe's stability and make its life more haz-

4

ardous: It is easier for a giraffe to break a leg than an elephant.

The giraffe has other problems as well. Because of its added height, it needs to pump its blood to a higher altitude than a man. This means it must sustain higher blood pressure and maintain tougher blood vessels. The taller an animal, the greater is its possibility of having a stroke—a broken blood vessel in the brain.

The simple fact that increasing a single dimension of any living creature also mandates an increase in all of its dimensions means a dramatic adaptation for it. A factor of 10 increase in length means, correspondingly, a change of 1,000 in volume and an increase of 1,000 in weight. Thus, as we compare animals of different sizes we find quite different behaviors.

Size differences have their consequences even among humans. For example, take a jockey and a basketball player. The jockey, being smaller, probably suffers little from back pains even with all the jostling he gets from racing horses. The jockey's compact body is able to withstand the vibratory movements; he can even fall from a horse without serious injury. Now picture a 7-foot basketball player riding a horse. Inside his large body are comparatively large organs, which, when jostled, would be subject to more damage simply because their movements would take up more space. The tall basketball player is also at greater risk just running and jumping on his own court. His proportionately thinner ankles and leg bones are greatly stressed in play—which is why so many basketball players suffer leg problems at some time during their careers.

Falling and Air Resistance

A mouse has little to fear from a great fall while a human worries over a fall of just a few feet. The reason? Air resistance. A man weighing 200 pounds who falls from a great height reaches a terminal velocity of about 200 miles per hour if he curls up in a tight ball. By doing this he decreases his exposed body area to the relatively high wind he experiences. If, on the other hand, he stretches out, offering more body area to the air he falls through, he reaches a speed of only 100 mph. Thus, the amount of body area a falling body exposes to a resistant medium such as air or water determines the speed of falling.

A stretched-out man reaches the speed of 100 mph after only a few seconds of falling time, and he continues to fall at this same speed (called the terminal speed) until he reaches the ground or opens his parachute. At terminal speed the force of gravity is actually overcome and balanced out by the force of air resistance.

5

When the man hits the ground at terminal speed, his momentum is great because of both his large weight and his speed. The mouse, having a smaller volume and weight by a factor of approximately 1,000, has a smaller skin-surface area buffeting the wind. Following the dimensions rule, its skin surface is, however, a factor of only 100 smaller. This means it has 10 times as much surface to weight as the man. Consequently, it reaches terminal velocity at only 20 mph, and weighing only 1/1,000th of the man's weight— around 2/10ths of a pound, or nearly 4 ounces—its momentum upon ground impact is considerably reduced, by a factor of 10,000. The mouse gets up from a 10-mile fall and laughs.

The Physics of Wetness

But while the mouse has nothing to fear from a fall through the air, it does worry about taking a bath. A 200-pound man rising from his bath has a coating of water all over his skin. It is about a 1/50th of an inch thick, and since water weighs about 64 pounds per cubic foot and a man has about 20 square feet of skin, this water weighs about 2 pounds. No doubt when you get out of the tub dripping wet, you won't weigh two pounds more because a good deal of the water will have dripped off. The sheer weight of the water is more than the force of surface tension, which causes the water to adhere to your skin, can hold. But even if that much water were to adhere, it would account for an added weight of only 1 percent.

For the mouse it's a different matter. Having 1/1,000th of the man's weight and 1/100th of his skin-surface area, the mouse has only 2/10ths of a square foot of skin. Consequently, it will be wetted down by just 1/3 of an ounce of water. Since the mouse itself weighs only 3.2 ounces, the water weighs 6.25 percent that of the mouse's weight. It is as if a man were to emerge from the tub with a 13-pound net draped over his shoulders.

For a fly, wetness is even worse. After it has fallen into the toilet, the sopped fly carries several times its own weight in water. This puts quite a damper on the aerodynamics: The fly can't fly. So while insects have little to fear from free falling in air, they do not like to get wet. If a small insect attempting to get a drink from a pond gets doused, the surface tension of the water is much too strong to be broken by the relatively small amount of weight that water holds. The bug drowns. That is why many insects have long noses (proboscises) to suck up the water from a safe distance. The water beetle avoids the wetting problem by having a skin surface that is aquaphobic, that is, oily. Its surface is simply not wetted by water. The water beads on its back like drops on a shiny, waxed Cadillac.

Area, Surfaces, and the Physics of Breathing

Although we shall be looking in greater detail at the function of the lungs in chapters to come, it is useful to examine them in the spirit of the previous section. What determines the physical size and structure of a lung? A human being has roughly 1,000 square feet of lung surface. (Actually this figure is closer to 861 square feet, but we won't quibble.) Our typical man weighing in at 200 pounds and standing 6 feet in height (again, in round numbers only) is covered by only 20 square feet of skin, yet he encompasses 1,000 square feet of lung tissue simply in order to obtain oxygen from the air. Spread his lungs out on the ground and they would cover a room around 32 feet by 31 feet. His skin, on the other hand, would only be a 4-by-5-foot throw rug.

Now, this lung surface is quite convoluted and folded up into many nooks and crannies. Altogether the lungs occupy only about 6 quarts of space (think of the space occupied by 6 quarts of milk in your fridge), yet they are the organ in greatest contact with the outside world. Each breath of air is about a 1/2 quart in volume. Each breath contains 10^{22} molecules of air. Since there are only 10^{44} air molecules on the earth, each breath takes in $1/10^{22}$ of all the earth's air. This is an astonishing statistic. It means that each 1/2-quart of air holds, on the average, 1 molecule once breathed by any human centuries or millennia ago. Just think, right now you just breathed a molecule of air breathed by Jesus or Moses, or, for that matter, Sol the butcher who lived in Minsk 200 years ago. Even a panting dog breathes in Jesus.

Why does a human need so much lung area? To answer, let us look at the animals again. First of all, every cell in a living body needs oxygen in order to survive. Since cells take up space, they occupy volume. The larger the animal, the more cells it contains and the greater its need for oxygen. Since oxygen is a gas, it must cross a surface area or skin in order to enter the animal's body.

The rotifer, a simple microscopic worm, has a smooth skin and no lungs. Yet it breathes. All the oxygen it needs is absorbed through its skin by contact with oxygen-enriched water. It has an immense surface-to-volume ratio compared to that of a human. A typical microscopic animal is around 40 microinches in length (a microinch is 1/1,000,000th [one millionth] of an inch). Its surface area is roughly 20,000 square microinches, while its volume is about 270,000 cubic microinches. This means its weight is only about $1/10^{14}$ of a pound. Its surface-to-weight ratio is about 14,000 square feet per pound.

Compare that with a man having 20 square feet of surface skin covering 200 pounds (a surface-to-weight ratio of only 1/10th of a square foot per pound). The rotifer is 140,000 times as efficient in using its skin-to-weight ratio to interact with its environment. It, correspondingly, has lots of surface around it and needs no additional convoluted surfaces of skin to take oxygen in. It has no need for a lung and wouldn't know what to do with one if it had it.

The man is a horse of a different color. In order to supply his great mass of tissue with oxygen, he needs extra square yards of breathing skin. His lungs provide just what he needs.

CHAPTER TWO

Them Bones, Them Bones, and What They Do

Going back to our giant once again, we found that a 60-foot-tall man would not really look like a man because he would need wider ankles just to stand up—to increase the number of square inches of cross-sectional bone area, and to reduce the compressional pressure on his ankle and leg bones.

Bone Dimensions

How big must a bone be in order to hold up weight? What about other forces acting on bones such as twisting (as in "I twisted my ankle") or shearing (as often occurs in football injuries)? For what other reasons are bones used?

Human bones come in a wide variety of sizes and shapes. The smallest bones, called the ossicles, are found in the middle ear. Laid end to end, these three bones, which hook up to the eardrum, would just about cross the diameter of an American penny. Even more surprising, these bones reach full adult size even before birth: The tiny fetus hears in the womb.

The fields of both dentistry and orthopedic surgery are devoted to the study of human bones. Since bones are about 22 percent calcium, and since calcium has a heavier atomic nucleus than most of the other elements in the body (an atom of calcium weighs 40 times as much as an atom of hydrogen), bone absorbs X rays much better than the surrounding softer tissues of the body. That's why bones show up so well in X rays.

The field of physical anthropology would hardly exist if it wasn't for bones' ability to last for great periods of time. Anthropologists have unearthed bones and skeletons that have shed light on our shadowy origins. Bones not only help us draw conclusions about the appearance and habits of our forebears, they help us ground these conclusions in actual time. Because bones are

9

made of carbon (about 16 percent), they have a natural radio-activity. This occurs because carbon exists in two naturally oc-curring isotopes: carbon 12, and carbon 14, the latter being radio-active, with a half-life of 5,700 years. After a creature dies, the amount of naturally present carbon 14 in its bones decays, which enables scientists to date bones by the amount of carbon 14 remaining.

What Are Bones Made Of?

Bone is made up of two quite different materials plus water. One of these materials, called bone mineral, is inorganic, composing about 60 percent of the bone's weight and 40 percent of its volume. This bone mineral is made of crystals. Each crystal is made of molecules composed of 10 parts calcium, 6 parts phosphorus ox-ide, and 2 parts hydroxide. Thus each molecule contains 10 cal-cium atoms, 6 phosphorus atoms, 26 oxygen atoms, and 2 hydro-gen atoms. These crystals are rod-shaped with diameters of 200 to 700 nanometers (a nanometer is one-billionth of a meter) and lengths of 500 to 1,000 nanometers. (For comparison's sake, the wavelengths of ordinary light are about 400 to 700 nanometers.) Since these crystals are so tiny, they have enormous surface-to-weight ratios of about 50,000 square feet per pound.

Since bone mineral takes up about 6 percent of the total body weight and has a density of 178 pounds per cubic foot (there are about 12 pounds of bone mineral in a 200-pound man), your bone mineral spread out would cover an area of about 164,000 square feet, or an area 406 feet by 406 feet.

Bone mineral's enormous surface-to-weight ratio makes it an ideal structural material, one capable of being surrounded by water containing, in solution, many materials needed by the body. This enables the bones to interact quickly with chemicals in the blood.

The Physics of Bones

The bones of the human skeleton perform at least six functions in the body: support, locomotion, protection of organs, chemical storage, nourishment, and sound transmission (as in our middle ear). (Cameron, p. 38.) In some other creatures, bone is also used in sex. All male primates, with the exception of man, have penis bones; indeed, even the walrus and the raccoon enjoy this "extra lift."

Perhaps the most intriguing functions of bone are the holding up of the body and the supporting of its movement through space. Walking, standing, running, lifting, even just getting up out of bed would be impossible without bones to lift weight. A study of the physics of bone, therefore, should prove useful in understanding the forces, stresses, and pressures we put our bodies through in accomplishing these tasks.

As if an engineer with a grand blueprint were at work, our bones exhibit a magnificent design. Bones are joined together by hinges called articulations, commonly referred to as joints. The function of bone joints still is not completely understood. In order for a joint to work, a special low-friction lubrication must protect adjoining bones from wear. Between two bones one finds articular cartilage bathed in a special low-viscosity liquid called the synovial fluid. This fluid contains acid and mucous-sugar compounds which have enormous molecular weights (around 500,000 times the weight of a single hydrogen atom).

It is the interaction between the fluid and the cartilage that is still something of a mystery. You might expect the surface of the cartilage to be as smooth as a ball bearing, yet it appears quite rough. Its roughness may be likened to tiny convolutions, nooks and crannies like those of a sponge, which act to trap the fluid and provide a smooth overall surface. One theory is that the cartilage even acts like a sponge; if true, then when the joint is squeezed, as, for example, when it is put under stress, lubricating material in the form of threads gushes out. These strings of lubrication are pulled back in when the stress is relaxed.

In either case, it is the synovial fluid that is most amazing. In a seemingly bizarre, Frankenstein-type experiment described by Cameron and Skofronick in their book *Medical Physics* (Cameron, p. 56.)—which, by the way, was a valuable source for this and other sections of this book—a normal hip joint consisting of the head of the femur (the thigh bone) and its socket was extracted from a fresh cadaver and mounted upside down (so that the socket was on the bottom). A heavy weight hanging below the mount was connected to the femur, thus pressing it into the socket (joint). The weight could be adjusted to study the effect of varying loads.

The hanging weight acted like a macabre pendulum. By swinging it back and forth, one could observe the decay (i.e., the rate of decrease) in the amplitude of the swing with time. By observing the decay of the swing and varying the weight load, it was discovered that the synovial fluid—which, given the freshness of the joint, was well intact—had an extremely low coefficient of friction. This COF, which was independent of the load (between 20 and 200 pounds) and the amplitude of the swing, was found to be less than .01, much less than that of an ice skater's steel blade on ice (which is about .03, by the way).

To grasp this conceptually, imagine pulling a 100-pound weight along the road. If the road is rough, you may need to exert 100 pounds of tug to keep it moving. This corresponds to a COF of 1. If you need only a 10-pound tug, the COF is .1, while a 1-pound tug corresponds to a COF of .01, the COF found in the hip joint even when stressed by 200 pounds of swaying weight. Clearly, synovial fluid is amazingly smooth!

The most obvious function of bones is support. To carry out this function, bones need to be strong. Looking again at what bones are made of, we find that in addition to calcium crystals, bone is composed of water and an organic substance called collagen. When isolated, the collagen looks like a flexible chunk of soft rubber. It has some tensile strength but is easy to bend into a loop. You might therefore think that the strength of bones is due to the calcium content. Yet after collagen and water are removed (as, for example, in cremation), the remaining mineral crystals are quite fragile, easily crushed with the fingers. It is the bone minerals that are saved in the cremation urn.

Yet put together, collagen and calcium crystals make a surprisingly strong, quite porous material. To grasp how strong, physicists study how a typical bone responds to applied physical forces: stretching, bending, compression, twisting, or shearing. Skiing accidents in which a leg bone is broken tend to be either shear or twist breaks.

Healthy bone, when examined closely, also shows a variation in porosity. If we take a typical femur or upper leg bone (the bone connecting your hip to your knee), we see quite a variation in the number of pores. In general, there are two classifications depending on bone porosity: compact, and trabecular. Trabecular bone is composed of long, thin, spongy threads called trabeculae, which appear at the ends of a bone, while compact bone is solid, appearing in the middle or central shaft.

Being more porous, trabecular bone is much weaker than compact bone. Its weakness is due mainly to the presence of less bone per volume of space, in much the same way as a honeycomb is more delicate than a stack of paper.

Why would nature make bones this way? To optimize the force distribution on the ends of the bones. The lines of force which act on the bone ends—for example, near the knee and hip, in the case of the femur—follow the trabecular lines. This lessens the force per unit area in any one region of the bone and allows for flexibility when the joints are put under extra stress. In other words, the bone ends are spongy enough to absorb stress. Furthermore, the trabecular lines guiding the lines of force are cross-linked by perpendicular trabeculae, which both reinforce and add springiness to the bone end.

There are two advantages, therefore, to trabecular bones: force distribution under compressive forces, and springiness under stress. On the other hand, since there is less bone per volume, trabecular bone cannot stand up to bending stresses the way the compact middle of bones can.

Here, in the central shaft of the bone, nature has utilized another principle of physics: economy. If you place a solid bar between two supports (as with a trapeze, for example) and hang a weight in the center of the bar, the bar will bend under the stress. The amount of bend will, of course, depend on the amount of the stress. This sets up what physicists call a stress-strain relationship.

In examining the bar, we learn that the stress is not uniform, but changes, at the place where the weight hangs, from compression at the top of the bar to tension at the bottom. The weight, in other words, causes the bar to squeeze together at the top and pull apart at the bottom. That means that in the middle of the bar, at the core, there is no stress, no compression, no tension. Engineers take advantage of this fact by using I-beams instead of solid rectangular bars as horizontal supports in buildings. If, however, bending forces are expected to appear from any direction, a cylindrical, tubular beam would be needed—and that is just what nature has provided in our bones: The core of our bones is hollow.

A hollow cylinder provides maximum strength with minimum weight for forces applied perpendicular to its length. What about forces applied along its length? If you push down on the end of a soda straw, it will buckle in the middle; so too our bones under a similar stress. However, the femur bone has extra thickness in the middle resulting in a narrower hollow; toward the ends it has less thickness and a larger hollow space. Thanks to this grand design, the bone is able to resist buckling under stress applied along its length.

CHAPTER THREE

How Your Bones Hold You Up and Sometimes Let You Down

Bones, as I have pointed out, are composed of collagen and mineral. It is perhaps surprising that this combination of materials shows so much strength. Yet together, they provide a material that is as strong as granite under compression, and 25 times stronger than granite under tension. Physicists have determined this by measuring how much a bone lengthens or compresses under the action of a force load. A similar measurement is made also by twisting the bone to check its strength under shear forces.

How Your Skeleton Works to Hold You Up

All materials change in length when placed under tension or compression. According to Hooke's law of physics, the relative change in the length, L (that is, the change δ-L divided by L) is proportional to the applied force divided by the cross-sectional area over which the force acts.

The force divided by the area is called the stress, and the relative-length change is called the strain. By dividing the stress by the strain, a ratio is obtained called Young's modulus, or Y. This ratio, for small-enough applied forces, remains constant (that is, it doesn't change in magnitude) as the force increases, and it is said to be a characteristic of the material substance. If the force applied is not too great, Y is the same for both compressive and tensile (stretching) forces. Whether you push or pull on the end of the bone, in other words, its Young's modulus remains the same.

The larger the Y for a material, the greater its resistance to change. Hard steel has a Y of 30 million pounds per square inch, while soft rubber has a Y of only 145 pounds per square inch. Granite's Y is 7.5 million pounds per square inch, concrete's, 2.4 million.

Compact bone and trabecular bone have vastly different Ys.

Compact bone resembles concrete with a Y of 2.6 million pounds per square inch, while trabecular bone has a Y of only 11,000 pounds. Knowing this, it is possible to calculate how much your body shortens when you stand up. (Did you know that you were taller lying down?)

Assuming that we disregard the change in length of your spinal column and look only at the length of your leg bone (which is about 27 inches long for the average adult), our 200-pound man, supporting his weight on the compact bone of one leg, shortens by only 7/1,000th of an inch. Your shortening would vary somewhat depending on your weight and height, but would be close to this number.

The trabecular region of the bone, on the other hand, would be compressed proportionally greater because it has a smaller Y. However, since the trabecular region is also shorter in length and wider in cross section (the ends of the bone are wider than the middle), the shortening comes to around 1/10th of an inch while standing on one leg.

Even though a large Y implies resistance to change, it doesn't necessarily mean high resistance to breaking or collapsing. Returning to the physicist's laboratory for a moment, we note that when the force applied to the material being tested is first increased, its Y does not change. However, as the force increases beyond a certain point, the Y begins to decrease at first slowly but then rapidly until finally a point is reached when Y is zero. The stress at this point is so much that even a small increase causes a whopping change in length. When that happens, the breaking point is reached and the material snaps in two.

There is a difference here between compressional breaking and tensile breaking of a material. Hard steel, for example, will crunch under a compressive load of 80,000 pounds per square inch, but it won't snap until a tensile load of nearly 120,000 pounds per square inch is applied. Granite crunches under 21,000 pounds per square inch, but will snap under a load of only 696 pounds per square inch. For concrete, 3,000 pounds per square inch for crunching, 304 pounds for snapping. Now you know why there are weighing stations for trucks. Since concrete roads crunch into dust whenever the load exceeds 1½ tons per square inch, heavy trucks must have extra wheels. This increases the number of square inches (footage) under the truck load, thereby lessening the stress on the concrete.

Compact bone is much stronger than concrete; it resembles granite when it crunches (at 24,650 pounds per square inch), but it is 25 times stronger under a tensile load (17,400 pounds per square inch). Trabecular bone is difficult to test under snapping conditions, easier with crunching—it gives way at 319 pounds per square inch.

Knowing the Young's moduli and the crunching and snapping stresses, we can determine how your skeleton holds you up safely and securely. Engineers like to build in a safety factor of 10 in their designs of structures; that is, they want the material they use to be able to withstand a stress load up to but not exceeding ten times the maximum expected stress.

Let's see how well your bones measure up to engineering standards under a variety of stress loads. Say you weigh 150 pounds. Whenever you walk, each step momentarily puts 200 pounds of weight on each leg. That is because your foot strikes the ground with some acceleration. Following Newton's law of action and reaction, the ground pushes up on each foot with an equal and opposite compressing force. During a high jump, this reaction (and your push down on the ground) increases to about 600 pounds.

Yet even this much force, with the wide area of the foot being 30 square inches, produces a stress of only 20 pounds per square inch. If you land incorrectly and put all 600 pounds on the heel of your foot, thus sending and distributing all 600 pounds to your tibia bone (your shin bone, which is about .4 square inch in cross section), you would place a stress of 1,500 pounds per square inch on the tibia. Since the bone crunches under 24,650 pounds per square inch, the safety factor is greater than 16 for running and jumping.

What about surviving a fall? Suppose you jump from a window at a height of 10 feet. When you hit the ground, you would be moving with a speed of about 17 mph. If you fell from 20 feet, you would hit the ground with a speed of 24 mph, while a fall from a height of 1,000 feet would mean an impact speed of over 170 mph. A fall from even greater heights would not increase your speed much over 200 mph because of air resistance.

The crucial consideration as far as breaking your bones is concerned is how long your leg remains in contact with the ground when it hits. The shorter the time, the greater the disaster. The reason is that the force transmitted up your tibia is, following Newton's second law, the change in the momentum divided by the time over which that change takes place.

In falling 10 feet and acquiring a speed of 17 mph, your body (assuming you weigh 150 pounds) possesses a momentum of around 117 pound-seconds. This number in itself does not say much, but it does express a potential for quite a jarring jolt, how-

16

ever. If your body were to come to a complete stop in about 1/10th of a second, it would feel as if it weighed over 1,000 pounds!

If you bounce on impact (called an elastic collision), your body suffers a momentum change of twice that amount, or 234 pound-seconds. Assuming your bounce took the same amount of time, 1/10th of a second, you would now feel as if you weighed over a ton. Your internal organs, of course, would be jostled by such changes in your body's momentum. When a boxer is struck in the head by a full blow, his head moves backward, but his brain, at first, does not. It is the change in momentum of the boxer's brain as it impacts against the inside of his skull that is so damaging. Similarly, your internal organs would suffer enormous damage.

If, however, you roll and absorb the impact (called an inelastic collision), your body will suffer a change of only 117 pound-seconds. Dividing this number by the number of seconds of impact determines how many pounds of force you will sustain on impact.

Typical impact times can be as low as 3/1,000ths of a second (3 milliseconds). If you landed on your feet from a fall of just 10 feet and attempted a stiff-legged or bouncy elastic-collision landing, which would minimize your impact time to about 3 milliseconds, you would experience a force of nearly 80,000 pounds. Distributing this over .4 square inch of tibia places a stress of 200,000 pounds per square inch compression on your lower leg bone. Since compact bone crunches under a load of 24,650 pounds per square inch, your poor tibia will not withstand this stress. It is no wonder that parachutists are told to roll when they land. A stiff-legged bounce will break the tibia unmercifully. A fall from an airplane without a parachute would increase this stress to over 2,000,000 pounds per square inch. For such a fall, a stiff-legged bounce will turn the tibia into dust.

On the other hand, if you allow your body to roll after a 10-foot fall, compressing upon impact and increasing the contact time to around 1/10th of a second, you would suffer an impact force of only 1,170 pounds. Dividing by the area of the tibia (.4 square inch) puts only about 3,000 pounds per square inch stress on the tibia bones of the legs. This stress is much less than the crunching stress of 24,650 pounds. Consequently, you will be able to get up and walk away from the fall.

If you could manage the same type of rolling movement after falling from an airplane, it would still produce 30,000 pounds per square inch on the tibiae—enough to break your legs.

So how good are our bones for taking the stresses of everyday life? They are remarkable! Even a fall from an airplane without a parachute would produce only minor tibia breakage if the faller rolls on impact. The design of our structural support system is quite adequate to handle the movements of everyday life, even if those movements require copious amounts of high jumping. It would be difficult, in fact, to come up with a better material than bone itself. Bone is the living scaffolding of human and other animal life. God and nature have worked together to provide high engineering tolerances indeed.

CHAPTER FOUR

Astronauts, Skeletons, and Menopausal Women

The Astronaut and the Skeleton

As the field of space travel grows, more and more people will experience the feeling of free fall during outer-space orbiting. Under such conditions, gravity is completely neutralized by centrifugal force.

A drive in the family car will illustrate what I mean. Whenever you come to a turn in the road, notice how your body responds to that turn. It will tend to move in the direction opposite to the turn. Thus, if you are making a fast right turn, your body will move, even sliding to the left. You are experiencing, from your point of view in the car, a force counter to the direction of the turn. This is the centrifugal force. In many modern chemical laboratories, devices called centrifuges are used; they rotate liquids containing suspended particles. It is the centrifugal force that enables the particles to press against the sides of the rotating cylinder, and, ultimately, to separate from the liquid that originally held them in suspension.

An astronaut inside a space satellite's room finds himself rotating around the earth. Consequently, he, like the passenger moving oppositely in a turning car, would find himself banging into the satellite's ceiling. He does not because of the oppositely directed force of gravity. The forces cancel out, and the astronaut floats freely.

This cancellation of forces takes place for every object inside the satellite. Thus, all the objects float freely. The human being and all objects regardless of weight or size become weightless. We might speculate what effect this would have on human skeletons if we were allowed to evolve in a weightless environment. The skeleton was designed to hold the body up under the forces of gravity. What shape would we assume without it? Would we grow long and thin? Or become rounded blobs with thin armlike appendages?

Actually, we can make a reasonable guess based upon observations of animals here on earth. We need not look up in the air to find the answer, but under the surface of our oceans. Because of the buoyancy of water, all submerged animals experience an upward force caused by the animal's displacement of water. Where the animal is, water isn't. The force buoying the animal upward equals exactly the weight of the missing water.

Fish, having densities very near to that of water, consequently experience little gravity when they swim; thus, they do not need strong skeletons. The lowly shark provides a good example because it has no bony skeleton—it's made of cartilage. Without gravity, the shark needs little in the way of bones.

In the absence of gravity, the geometry and shape of a body will change. The large water mammals, such as dolphins and whales, provide a picture of what we might begin to resemble floating in outer space. As stresses due to weight disappear, we would become more rounded and spherical. Our legs would either disappear or adapt to become handlike limbs similar to those of monkeys. Our arms would become long and thin as we found the need to reach out to the surrounding surfaces or supporting bars of our spacecraft to anchor ourselves from free fall, spinning, and gyrating.

There is already firm evidence that humans alter their skeletons in outer space. (Leach, Wronski.) Research with astronauts has shown that the calcium absorption decreased in orbit, while the calcium excreted in the urine increased. This indicated a decay in astral bone matter. What does the calcium loss indicate? Possibly, according to Dr. T. J. Wronski, bones in space begin to lose mass and strength simply because they are not needed.

Since bone is living tissue containing a blood supply as well as nerves, it undergoes continual change throughout life. A special class of bone cells carries on a continuous process of destruction and creation. *Osteoclasts*, or bone eaters, destroy cells and *osteoblasts*, or bone growers, build them.

Building bones is called bone remodeling. However, the race to build before destroying is quite uneven. As in many physical processes, destruction is always more efficient than construction. One bone-eater cell can destroy bone cells about 100 times faster than one bone grower can create them. It takes about 7 years to completely regrow a new skeleton. But each day about 1/1,000th of a pound of bone calcium is "eaten" by osteoclasts. When you are young and growing, the bone growers outdo the bone eaters (that's why children need more calcium than adults); but as you age past 40 years, the bone eaters begin to win the race. As you approach old age, your bone mass slowly decays—a process that continues until you die.

Menopausal Women and Soft Brittle Bones

It is a well-known fact that some women lose bone calcium during their childbearing years. During pregnancy, the growing fetus will rob its mother's bones and teeth if she fails to take in enough calcium in her food. Yet for some unknown reason, even childless women lose more calcium with advancing age than men. This can lead to a serious disease called osteoporosis (which means porous bones). According to a 1984 Scripps Clinic and Research Foundation report, this disease affects 7 million postmenopausal women in the United States and exists in a nonsymptomatic form in another 7 million. Osteoporosis, which appears to be a metabolic disorder, can cause a progressive loss in height of as much as 6 to 9 inches. It is often accompanied by the characteristic spinal curvature known as dowager's hump.

George E. Dailey, III, M.D., head of the division of diabetes, endocrinology, and nephrology at the Scripps clinic in La Jolla, California, explains: "Bone loss occurs very gradually, at an approximate rate of 1 percent of the skeleton per year. This is not measured by standard X-ray techniques, which do not reveal a change until almost 30 to 50 percent of skeletal minerals have been lost."

With the loss of calcium, bones become more porous and therefore not only softer but also more brittle. A typical injury suffered by elderly people is a broken hip. Each year about 150,000 women in the United States sustain broken hips, and usually these women are suffering from osteoporosis. Why the hip bone in particular? While all bones would suffer from the disorder, the calcium loss would tend to be more severe in this region due to the greater amount of stress it receives in walking. Remember that under stress, bone will wear down. A normal bone will regrow, but the diseased hip bone cannot repair itself. One day, sometimes under a relatively minor stress such as stepping off a curb, it simply gives way.

Although the basic cause of osteoporosis is not known, a new diagnostic tool for early recognition of the disease is available. A dual photon beam absorption densitometer is now being used at the Scripps clinic to detect minute changes in bone density. A beam of high-energy photons (X rays) in small doses is absorbed by the bone. From the absorption the amount of bone calcium is determined.

To prevent the disease, sufficient calcium in the diet is required (about 1,000 to 1,500 milligrams are considered a daily minimum). A single glass of milk contains about 250 grams of calcium. (Note that a gram is quite small. To make up just 1 ounce you need a little over 28 grams.) Also, activity or exercise appears to help. By moving the body and putting stresses on the bone, the skeleton continues to develop instead of deteriorating.

Thus, it appears that bone growth and regeneration do depend on forces we exert on our bodies during typical movements. Weight-bearing exercises such as walking, running, and bicycling appear to strengthen bones, while non-weight-bearing exercises, such as swimming, may not. If the evidence of outer-space activity is conclusive, then humans will need to consider artificial means for ensuring that bone development continues in space. This would be a problem for young children raised in outer space.

CHAPTER FIVE

The Quantum Physics of Fitness and Fatigue

How does the body use energy to move itself from place to place and to keep its internal organs in motion. The key word here is "energy." Without energy the body will not and cannot move or function at all.

The Energy That Moves Us

Energy, as we have noted, is a rather mysterious term. It means, in physics, no more than the ability to do work. It can be measured in many ways. Even though the different ways of measuring energy are quite varied, they all boil down to the same basic thing, best known by the public at large as calories. A calorie is thus a unit of energy measure.

The body requires calories. Just how many depends on the kinds of processes the body is engaged in. In this chapter we will examine some fundamental processes of converting energy from food or stored energy into movement or kinetic energy.

A few key terms will be needed here. Any process that converts food energy into movement energy is called *glycolysis*. The food that is converted is glucose or blood sugar. If glycolysis takes place in the presence of oxygen, it is called *aerobic glycolysis*, or simply oxidation. If glycolysis occurs in the absence of oxygen it is called *anaerobic glycolysis*.

Why Do I Get Tired When I Work Hard?

We've all been through it. After a long day at the office or after a particularly stressful situation, we feel tired. But what causes the tiredness we feel? Lately, we see joggers abounding around us with what appears to be all kinds of energy. In the 1960s, hardly anyone ran any distance unless trying to catch a bus or in training

23

for the football or basketball team; through the 1970s and into today, many of us are out there first thing in the morning jogging away one, two, three, or more miles, two, three, or more times a week.

What enables the runner to keep going even as far as 26 miles without stopping for rest, food, or air? How is it that some days, particularly the unstressful ones, I come home from the office feeling energized and ready for a full evening, and other days, I want to collapse in front of the TV?

Part of the answer to the above is now known. I propose that within this answer is a pathway for consciousness to enter which will someday answer the question completely.

The main reason we get tired is lactic-acid buildup in our muscle cells. Lactic acid is the product of anaerobic glycolysis, a process occurring in all cells if the supply of oxygen is limited for any reason. In this case, the glucose in the muscle cells is not allowed to produce the most energetically favorable products. Instead, glucose is converted to lactic acid, quickly depleting the glucose and exhausting the energy supply for the cell. We'll look at some of the details a little further on in this chapter.

When oxygen is present—normally the case and why we breathe in the first place—a series of chemical reactions takes place in the body cells. These reactions involve many complex molecules including enzymes (about which I'll explain more later). These reactions occur in most aerobic organisms, especially during respiration, and are collectively called the *Krebs cycle*.

The Krebs cycle is a remarkable series of aerobic chemical reactions occurring in every cell of your body. It is the major pathway by which metabolism (energy conversion) in the presence of oxygen takes place. The goal of the Krebs cycle is to produce a certain highly energized molecule known as *adenosine triphosphate*, or ATP for short.

The Krebs cycle is necessary to make energized electrons, which in giving up their energy ultimately produce ATP. The reason we need oxygen is that these electrons in producing ATP must land on a lower-energy site. Thus, they finally make their way to oxygen O, forming negatively charged oxygen ions, which are attracted to protons, H^+, to then make water, H_2O.

For an electron, oxygen is a powerful attractant. By analogy, the electron making its way to oxygen is like a ball rolling along until it comes to a precipice. The fall of the ball down the steep decline gives it kinetic energy. In a similar way the fall of the electron into an oxygen atom gives the electron energy which it, in turn, uses to produce ATP.

If oxygen is not present, the attempt to make acetate from glucose fails. Instead lactic acid is made, and making it is expensive for the body. Only a small amount of energy is made available.

If we compare this with the oxidation of glucose—the burning of glucose in the presence of oxygen—we find a much greater production of calories. The burning of glucose produces 13 times more energy compared with its fermentation, which results in lactic acid. In an aerobic process, when lactic acid is not produced, other mid-products, called *pyruvates*, are oxidized with carbon dioxide and water as the ultimate end-products.

The key insight here is that the amount of energy produced from one molecule of glucose is much higher via oxidation than via anaerobic fermentation. If you consider your body as a well-timed machine, supplying your cells with glucose from your blood at a certain rate, then it is easy to see that anaerobic fermentation, in consuming a lot of glucose in comparison to aerobic burning, would soon exhaust the supply. Demand outweighs supply and the cell goes into shock.

This shock is easily experienced as tiredness. It is quite common for athletes to experience this when they exercise their muscles to the limit. This limit is nothing more than reaching the point where anaerobic processes are matched by aerobic processes in the muscle cells. The athlete, by using his or her muscles to an extreme, is forcing the muscle cell to move and respond more quickly than the blood is able to supply glucose to the cell.

The tiredness you feel after a long day grinding it out at the office is probably not the same as the tiredness felt by a sprinter after a 100-yard dash. For an office worker under stress, tiredness is probably equally due to muscle cramping, resulting in poorer blood supply, and possibly poorer nerve communication. The exhausted office worker's muscle cells are indeed hungry because of such tensions. The sprinter's muscles are also hungry, but not because of tension: Her muscle cells are simply exhausted of glucose because she demanded too much from them and anaerobically consumed her supply. Not enough oxygen was able to reach her cells in that short 10-second dash for the finish. Lactic acid was made instead of the necessary pyruvates.

Anaerobic glycolysis does produce some energy: 2 molecules of ATP. However, aerobic oxidation of glucose produces 36 molecules of ATP for each glucose molecule consumed. With strain produced by overexertion and mental stress, muscles rapidly use up their chains of glucose (called glycogen).

The ability to overconsume without oxygen must, however, be

very important—otherwise we wouldn't have developed this pathway for energy production. Perhaps this knack harkens back to our primeval beginnings as cells. The primal cells, which probably developed before oxygen was even present in the atmosphere, would have had to use anaerobic glycolysis in order to live. Their need for food would have been enormous. With the creation of oxygen in the atmosphere and its subsequent appearance as a solute in water (that's how fish breathe), the more efficient aerobic oxidation of glucose became possible, supplying over 36 percent of its energy to form ATP.

Evolution eventually brought forth more complicated creatures, and with them more violent times. Fight, flee, or fornicate became the order of the day. To survive, a prehistoric animal needed to be able to move quickly—faster than its aerobic glycolysis processes would allow. The earlier anaerobic glycolysis did the job. The animal could run away without taking a breath. Of course, it eventually had to breathe, but its ability to last without oxygen became a measure of its fitness for survival. By being fit it had both plenty of glucose around and a more elaborate network for supplying oxygen to its muscles when oxygen was available.

If there were plenty of glycogen around, and plenty of it supplied as glucose units from the liver where glycogen is manufactured, humans would have no need for breathing. Living without oxygen for long periods of time in the manner of whales and other seafaring mammals would be entirely possible. The reason we humans die after only a few minutes of oxygen starvation is a lack of glucose in the cells. Since 36 ATPs are produced per glucose oxidation as compared to only 2 molecules of ATP from anaerobic glycolysis, 18 times as much glucose needs to be consumed in order to produce the same amount of ATP.

Muscles: Where the Mind-Body Interaction Takes Place

Tiredness from physical exertion and mental stress is probably felt more keenly in the muscles. And, as we have seen, by tiredness we mean a deficiency of energy produced by inadequate ATP molecules to meet the energy requirements of the body cells. Tiredness also occurs when oxygen is not supplied in large enough doses to carry out the oxidation of glucose. Instead, anaerobic glycolysis takes place, producing lactic acid that uses up the glucose 18 times as fast as when oxygen is present in the cell.

Muscle cells are quite complex and, I believe, extremely important to grasping how the mind-body interaction changes and alters the body. Muscles are able to maintain profiles. They may become tensed as a result of thinking about something troublesome, and then that tension may linger. Even your sense of self may "reside" in your muscle cells. Enough tense muscles can create a "body armor" as means of protection—a kind of externalized ego.

Thus, muscles provide an arena for both classical bioenergetic processes and mental processes. Muscles are the externalization of the mind.

To grasp how the mind-muscle interaction occurs, we must first look at the bioenergetics of muscles and how muscles are constructed. Muscle makes up about 35 to 45 percent of the body's weight. It exists in three forms: smooth muscles, cardiac muscles, and skeletal muscles.

Smooth muscles are made of elongated cells arranged in bundles. Each cell has a single nucleus. They are found in the intestines in ringlike bands able to produce wavelike contractions controlled by the autonomic nervous system (ANS). They are also found in the arteries and the urinary tract. These muscles contract involuntarily.

Cardiac muscles forming the heart are made of branched cells, each containing several nuclei. They occur in spiral bands around the ventricles of the heart and are able to perform rhythmic contractions under the action of the heart's pacemaker—again, an involuntary process. What is absolutely amazing about cardiac muscles is that they never tire out or need rest unless an infarction (cut-off of oxygen to the cells) occurs.

Skeletal muscles are under our conscious control. Each muscle consists of elongated cells, each up to as much as one foot in length, and each containing many nuclei and many minute fibers called *myofibrils*. We are consciously able to control the bioenergetics of muscle contraction and relaxation, and when we do so calcium ions are released. Skeletal muscles represent the ideal mechanism exhibiting the mind-body interaction.

It is also in skeletal muscles where tiredness and stress are clearly manifest and responsible for our feelings of well-being and health. Weakened muscles unable to support the joy of spontaneous body movement are greatly responsible for depression. This weakening of muscles is brought on by the inappropriate utilization of glucose and oxygen, the chief ingredients for muscle health and well-being.

A Proposed Scenario of How a Voluntary Muscle Contracts

The process of muscle contraction and relaxation is probably the most important aspect of human behavior known today. Through these actions we get visual clues from each other, communicating our true feelings even when our words lie. Without these actions we wouldn't be able to move or express ourselves. Every facial expression of disgust, desire, excitement—"Wow, that was an eye-opener"—every gesture, the latest rock 'n' roll dance, the ability to speak, stand, walk tall, slump, even sleep, require conscious actions directed to and responsive of the skeletal or voluntary muscles.

Without skeletal muscles the body would be floppy and incapable of any movement at all. Each human being possesses over 600 voluntary muscles. They are so named because these muscles can be consciously controlled. They rank in size from the bulky gluteus maxima, or buttock muscles, which consist of many layers of thousands of myofibrils, to the tiny stapedius muscles just inside the cavity of the middle ear.

Muscles join onto each other. They are said to have an origin at one site and an insertion at another. At the origin, muscle fibers are attached directly to connective tissue covering bone. At the insertion, the muscle usually tapers off, its fibers replaced by cablelike tendons that are relatively inelastic and thus incapable of stretching or contracting. Sometimes the muscles are inserted into the fibrous sheaths surrounding other muscles.

Muscles usually work in opposition to one another. One muscle contracts while another stretches. A common example is the muscles comprising the upper arm known as the biceps and the triceps. When the "front" biceps contract, the "rear" triceps relax. When the arm is straightened, the opposite action occurs.

INSIDE THE SKELETAL MUSCLE CELL

A closeup look at a muscle reveals just how muscles contract under conscious direction. A single cell is long and cylindrical. Its many nuclei provide an ample supply of nucleic acids for templates used for protein synthesis and other constituent buildup. Its myofibrils are the basic functioning units under conscious control and are able to contract and relax voluntarily.

When seen under an electron microscope, each muscle cell and each myofibril within the cell shows a cross-striated appearance. That is because they consist of regularly occurring alternating zones exhibiting sharply different optical properties. In active muscles, mitochondria are usually found adjacent to the cross-striations in regular arrays; they are the energy producers which convert the ATP to ADP to provide energy for contractions.

A closeup look at these zones shows they follow a repeating pattern. Each section of the myofibril consists of repeating units arranged lengthwise along the long axis of the cylindrical muscle. These units are called *sarcomeres*.

There are two basic parts to a sarcomere: the thin filament, and the thick filament. The thin filament, called *fibrous actin* or *F-actin*, is made of a long double chain of molecules tightly spiraling around each other like two snakes in love.

The thick filament is made of myosin molecules. These consist of thick strands of polypeptide chains with loose barblike heads sticking out. The heads contain short polypeptide chains and an enzyme called ATPase which acts to convert ATP into another molecule called ADP. This uses the energy of ATP to contract the fibers composing the thin filament.

MOLECULAR BRIDGES

In order to make a voluntary muscle respond by contraction or stretching, a sequence of steps involving the making and breaking of molecular bridges takes place. The process is at once quantum physical and classical mechanical. The thick muscle filament is made up of many stringy molecules winding around in parallel strands, like the strands of a strong rope wrapped around each other. Sticking out from the main body of the "rope" are the tiny projections consisting of the heads of the long myosin "strings." These tiny heads, actually short polypeptide chains, contain the enzyme ATPase enabling ATP to be converted into ADP at the head.

The thin muscle filament is actually made of ball-like molecular units called G-actins—globular proteins with a molecular weight of around 46,000. These balls of G-actin stick together, making a double strand of recurring G-actins, the F-actin strands making up the thin filaments of all muscles.

Actomyosin (actin and myosin together) is formed when the thin filaments and the thick filaments link cross-bridges between them. It is the alternation between cross-bridge building and breaking that actually accounts for the contraction of the muscle.

When a muscle contraction takes place, the thin filaments are moved or pulled into smaller sarcomeres. The heads of the molecules are able to perform a kind of ratchet and pawl action together, provided the catch and release takes place fast enough. When the muscle relaxes and elongates once again, the ratchetlike heads of the myosin molecules are released from the thin filaments and retract back toward the thick filament. This allows the thin filament to slide over the thick filament without impedance and return to a relaxed state. In the relaxed state the thin filaments are elongated.

During contraction, doubly ionized calcium ions CA^{++} rush into the space between the F-actin (thin filaments) and the myosin (heads on the thick filament). When this happens the heads of the myosin molecules are attracted to sites on the F-actin molecule. This causes a set of molecular bridges to be erected, forming the actomyosin complex which, in effect, gives muscle its rigidity and its tension. The cross-bridges of myosin that inappropriately remain after the command to release is given provide the major cause of headaches, backaches, and muscle pain in general.

If the process uses up the calcium ions, the bridges remain and the muscle can appear rigid. In fact, when the muscle is relaxed in the absence of free calcium ions, the cross-bridges will remain intact with the sarcomere volume as large as possible. I don't want to give the impression that cross-bridges are the cause of tension in and of themselves. It is the inappropriate cross-bridging which remains—due to a lack of free calcium ions—when the sarcomere volume is contracted that results in that old tired feeling and headaches.

With a further influx of calcium ions, the cross-bridges lose their footholds in the actin and myosin and the bridges break. These ions stimulate ATP to release its energy, and this action breaks the bridging bond. With bridges broken, the actin and myosin can again slide over each other to the relaxed state. As long as calcium ions are present in the actomyosin space, bridges will continue to break and reform repeatedly, each break and reformation using up ATP energy. It is the repeated break and release action that is the cause of muscle contraction.

When free calcium ions are no longer present, the muscle gradually relaxes as the sarcomere elongates. Thus, the muscle relaxes again when the calcium ions rush out of the space inhibiting the reaction ATP $\rightarrow$ ADP.

Consciousness and Calcium Mediation

Calcium mediates the contraction and relaxation of all muscles. It is also true that another molecule called *phosphocreatine* can play a role in supplying energy in the absence of ATP. However, the main source still appears to be the ATP to ADP conversion mediated by calcium ions.

When a person undergoes some form of trauma, the body tends

to hold that trauma in the muscles as well as in the mind. Evidence of this is seen in body posture. Typically, newly released prisoners, after serving many years, show body postures indicating a whipped-dog look. Often this is compensated for by an exaggerated egoic chest puffing out. Children that are berated often walk with their heads down. The concept of body language, indicating how we really feel about unpleasant or secretive situations, is another form of evidence that we hold trauma in our bodies.

Some people who develop hearing problems reveal that they really don't want to hear what's going on. I myself remember the day I became nearsighted. As a child, I was horribly afraid of sitting in the rear of the room and being ignored because my last name began with W, and at the same time did not want to be called on to recite because I stammered badly. Nearsightedness solved my problem. By not seeing the board too well, I had to sit closer, and because I couldn't see the board, I was excused from reciting. Of course, I eventually had to get glasses.

In a similar vein, chronic backaches can plague people who feel they lack support. People with laryngitis may fear voicing their opinions or be repressing anger. People with leg problems often fear the future—they are afraid of walking toward it. If you can't "stomach something," meaning you can't accept it, you may suffer a gastrointestinal muscular spasm. And if you are physically crippled due to emotional "crippling," you may indeed be cured by the spirit of healing at a "Jesus Saves" revival.

If muscles are capable of "holding grudges," how does this happen? A grudge held is a muscle fixated in a certain position. This means that calcium ions are continually flooding the sarcomere. Once the trauma occurred, the calcium remained and the muscle was not allowed to relax; the tension was held by a continual breaking and making of the myosin bridges in the actomyosin molecules, thus causing the muscle to remain in the contracted position. This fixation took place in an appropriate body location. Thus, consciousness entered and locked in through the maintenance of calcium in the sarcomere. This "consciousness" could be construed as the Freudian unconscious. However, I want to posit that it is the same consciousness that enters in the observer effect of quantum physics mentioned earlier. What we call our minds may exist throughout our nervous systems and even in our muscle cells.

These muscle lock-ins can easily be felt in body work or body massage. Often a gifted healer working with a patient suffering from muscle pain or discomfort will mentally "flash" when he touches the afflicted area. The flash will reveal the actions causing the unrelieved constriction of the muscle. Many times the patient will suddenly realize thoughts related to the muscle cramp as the massage of the afflicted area takes place.

Reich's Orgone and the Qwiff

Probably the first psychologist to realize that muscular contraction and consciousness could lead to disfunctioning and inappropriate tiredness was Wilhelm Reich. Reich, who worked with the concept of bioenergy, believed that the body uses its muscles as a form of muscular armoring as a reaction to stress. His concept was:

Mechanical Tension → Bioenergetic Charge →
Bioenergetic Discharge → Mechanical Relaxation

Reich would not have been aware of the nature of calcium ions in this process, nor that the charge he referred to was a real electrical charge, the charge carried by the calcium ions themselves. We now know that when a muscle undergoes contraction, calcium ions bathe the sarcomeres causing contraction to occur, and, therefore, the bioenergetic discharge Reich referred to was the outflow of calcium ions from the regions of the sarcomeres.

All muscles at rest give off heat at a low and constant rate. This is called the *resting heat*. When a muscle is stimulated to contract, a relatively larger amount of heat is given off. This is called the *initial heat*. Some of this heat is given off just in the process of producing the tension before the actual contraction occurs. The remainder of the initial heat is produced during the actual process of contraction. When the muscle relaxes again, another bout of heat production takes place. This is called the *recovery heat*. Recovery heat is a measure of the electrical processes during the outflow of calcium ions and the recovery of ATP from ADP through the processes of oxidation.

The overall efficiency of muscle work has been determined to be only 20 percent. When a muscle is under continual tension, it follows that it will be producing more heat than when relaxed. Thus, it is possible to feel hot or warm spots on the body, indicating tension. Since the muscle is also burning hotter, it will tend to tire faster than the surrounding relaxed muscle tissue. It will consequently require more food input just to keep going and so will fuel its tension at the expense of energy needed for useful work by other muscles. In this manner, by keeping muscle cells hot, con-

32

sciousness is able to hold body memories as a defensive measure.

When a continuing circumstance occurs, such as a spouse yelling at the other spouse, the delicate ear muscles (the stapedius muscles) will remain contracted, with calcium ions causing the contraction in an attempt to lessen the noise. When an offensive football player is continually subjected to attack by an oncoming defensive front-line player, he will produce a steady contraction of neck and shoulder muscles in order to prevent bone and nerve damage.

Each of us carries some muscular armor in order to protect us from some outside "invasion." The problem is that once the muscle is trained to protect, it is not always easy to untrain it. A guard dog is not always the best pet.

Many people carry tension in their heads, necks, and shoulders. As children, such tension probably arose as protection from being hit by other children, or, sad to say, by parents. This tension is reinforced by desk work that requires that the upper torso be held in a restrictive position most of the work day. As a result, head and shoulder muscles are sore and in pain.

Reich based many of his observations of human beings on the belief that an invisible energy pervaded all of the universe. Reich was also influenced by Freud in that he believed that this energy was akin to Freud's theory of the libido. Thus, it had much to do with sexual functioning, according to Reich. From my point of view, Reich was probably on to something, in that there may be a connection between energy and quantum physics—a speculation I offered in my earlier book *Star Wave*. Certainly we know that having sex requires muscle cells operating in a cooperative manner, and the pleasure most of us feel during sex may be our tuning in to this "energy." Reich, in fact, called this energy by the name *orgone energy*—suggesting a connection with human orgasm.

He defined it as:

1. Mass free, having no inertia or weight;
2. Present everywhere (though not in equal amounts), even in empty space;
3. The medium for all electromagnetic and gravitational activity, the underlying basis for all movement;
4. In constant motion, and observable under special circumstances;
5. Capable of going against the law of entropy—high concentrations of orgone attract low concentrations of orgone; and,
6. Capable of forming units that include cells, plants, animals, and also clouds, planets, stars, and galaxies.

Reich was born in 1897. He made his controversial discoveries during a period when quantum physics was also being discovered. I believe that Reich's orgone may be the same as the quantum wave function of physics. Like orgone, the quantum wave function, or qwiff, is:

1. Mass free, having no inertia or weight;
2. Present everywhere (though not in equal amounts), even in empty space;
3. The medium for all electromagnetic and gravitational activity, the underlying basis for all movement;
4. In constant motion but unobservable except as a pattern of probability under all circumstances;
5. Capable of going against the law of entropy (qwiffs from interacting systems coalesce, forming correlated systems); and,
6. Capable of forming units that include cells, plants, and animals, and also clouds, planets, stars, and galaxies.

Following Reich's concepts of orgone energy, and inserting my idea that orgone is not really an energy but a quantum wave of probability, the body's muscular armor is caused by the mind's altering the probability patterns of calcium production and release in the muscles. Once this armoring effect occurs, once the muscle learns to continually produce calcium or maintain its presence in the sarcomeres, the conscious mind no longer plays a role.

This means that nerve impulses originally causing the inflow of calcium no longer have the effect of control. The muscle is, so to speak, an isolated pocket of consciousness, unable to respond to the nerve command to relax. The calcium ions simply do not leave.

If my concept is correct, body armor can be altered by thought. If you, at this moment, feel that a particular muscular group is quite tense, try to isolate that group in your mind. Indulge in some mental imaging. Picture the tense muscles as composed of many tiny rope bridges all held tightly by little soldiers. Now, gradually allow the soldiers to let down their ropes; as they do so, the bridge held by them begins to slacken and your muscle cells will begin to relax.

As the muscle relaxes, feel your energy begin to return as more of it becomes available to you, released from the chore of holding on to the rope drawbridges of your muscular armor. With each relaxation, more energy is being converted via aerobic glycolysis rather than the tension-producing anaerobic glycolysis, and more calories are being made available to refresh your brain. Your body will feel better, and since you are now using oxygen more efficiently (it helps if you breathe deeply while doing this), you may not need that extra fuel snack after all. By relaxing your muscles, you may find more blood sugar available since you aren't wasting the energy in lactic-acid buildup.

CHAPTER SIX

The Quantum Physics of Exercise and Stress

How do exercise and stress affect us? Up to now we have seen how classical physics enters the body human. Here we venture into unknown territory: the body quantum.

First we look at the question of health and fitness. Having seen that they are not synonymous, we attempt to make a model of stress and its release. I propose that stress is a form of information feedback from the nervous system to the muscles. By using a simple idea from quantum physics known as the observer effect, I hope to show how our consciousness enters into that feedback loop. Perhaps the ideas presented here may be helpful in learning to release that stress.

Fitness: A Sacred Activity?

In today's Western world view, body fitness has become a sacred activity. In the early 1960s, after finishing my Ph.D. thesis work, I found myself 20 pounds overweight and sadly out of shape. I decided then to take up running. I began a process of slow running and alternate walking, gradually increasing the duration of running, decreasing the walking, and increasing the distance. In a year I was fully running distances of two to three miles three or four times a week. When I took up this activity, there were virtually no runners on the streets; fitness centers were called gyms, and were frequented only by weight lifters, boxers, and professional athletes.

By the mid-1980s, the scene has completely changed. Now runners jam the streets, fitness magazines clog the newsstands, and fitness centers are drawing crowds. These days I too have added two sessions a week at the neighborhood Nautilus center to my routine.

In 30 years of using exercise regularly—as a youth I took part in football and basketball—I know that health and fitness are strongly increased by regular exercise. I have also noted that during those times when illness or work has kept me from exercise, my fitness declined. With advancing age (I am now 50), I see that it takes me longer to recover my fitness from lapses in regular exercise than it did when I was 10 years younger.

Fitness can be improved by two kinds of exercise: aerobic (meaning, "in the presence of oxygen") exercise, and anaerobic ("in the absence of oxygen") exercise. Anaerobic exercise is typically used to increase muscle bulk, while aerobic exercise is used to increase stamina. By anaerobic exercises, such as weight lifting, the muscle cells are quickly deprived of glucose. Since the increased rate of doing work brought on by the exercise causes them to become depleted, there is usually not enough oxygen supplied by the blood vessels feeding the muscle cells. In aerobic exercise, the muscles are never completely deprived of glucose, so that the blood supply of oxygen is adequate yet still being taxed by the muscle movement. An unfit body, in addition to having small or weak muscles, simply doesn't have an adequate blood supply to its muscles. Aerobic exercise appears to increase the blood supply to the muscles by actually producing more arterial blood vessels in the muscle "neighborhood." Anaerobic exercise appears to increase the muscle size and strength. The increased growth of blood vessels in aerobic exercise sprouts from the blood vessels closest to unfit muscle. However, there now appears to be some controversy about how exercise actually improves fitness or health.

Fitness Versus Health: A Controversy

We obviously know from experience that stressing a muscle increases the muscle's bulk and strength. We also know that lung capacity increases and the cardiovascular system improves with regular exercise. Yet, although an enormous amount of data has been accumulated concerning fitness and exercise, it is not clear just how exercise really increases fitness or produces health benefits, or why fitness or health declines from lack of exercise once a level of fitness has been reached.

The question of how fitness and health are related has led to some controversy.

There are two schools of thought. According to one, many fitness experts and trainers believe that high-intensity anaerobic exercise, such as the type in the Nautilus fitness program, is essential for maximum health and fitness. This means using equipment in brief and infrequent intervals to stress the muscles to their fullest capacity. Each exercise session is followed by a much longer period of rest. The result of such a program will be to produce a fitter muscle—one capable of performance at higher efficiency. The reason is due to the increase in blood supply to that muscle. Remembering the early days of my athletic career, I can verify that anaerobic sprints produce the desired results.

But, according to the other school, such intensive exercise may lead to greater fitness but poorer health. Whenever a muscle is stressed to its fullest over a short period of time, it must perform anaerobically (i.e., without oxygen; see previous section). This means that the muscle must use glucose (sugar) to produce lactic acid. Lactic acid, in turn, causes blood vessels to dilate (expand), increasing blood flow and making oxygen more available to the anaerobic muscles. This process coupled with increases in pulse rate and blood pressure is called "paying off the oxygen debt."

Since the circulatory system contains a finite amount of blood, the dilated blood supply to the stressed muscles may take the blood supply *away* from organs such as the kidneys, liver, and brain. Even after a period of high-intensity exercise, these tissues will be left in a state of oxygen-deprived shock until the blood vessels to the muscles contract once again.

To counter this deleterious effect, the second school, as typified by the Nordic fitness program, advocates a milder form of muscle stress by exercising below the lactic-acid-buildup point. In this manner, all the organs receive the benefit of the increased activity. By exercising slowly to begin with and gradually increasing activity and never getting "winded," one will not increase any particular muscle mass or fitness but will produce an overall health benefit. The idea is to avoid strenuous exercise—to not create in any muscle a lactic-acid buildup.

How does the fitness-conscious person make sense of all the contradictory information? My advice is to concentrate on aerobic exercise, making sure you increase blood circulation gradually during each session. Allot about 15 minutes to warm up by performing the particular exercise at a slow pace. For example, begin a running session by walking slowly and then increase your speed very gradually over the 15 minutes. If you also want to increase muscle mass and fitness, my advice is go ahead and work with the free weights and the high-tech contraptions—but do some aerobic exercise before each workout. This will open up the blood supply to the whole body before the taxation of oxygen needed for muscle fitness occurs.

A Feedback Model of Stress

If we examine the simplest aspect of muscular stress and release, we come back to the fact that muscle cells are composed of units which undergo changes in length and internal tension. Muscular elongation or contraction is accompanied by either muscular stress or stress release. In other words, the cellular unit is either stretched under tension or stretched with tension release, or it is contracted under tension, contracted with release.

Thus, the relief of tension can be accompanied by either an elongation or a contraction of a muscle unit. Under normal or so-called involuntary muscle control, outside forces act on the muscle cell unit, putting it under tension. When the tension decreases momentarily, the muscle unit elongates; when the tension momentarily increases, the muscle cell unit contracts. This process occurs at the molecular level, most probably mediated by calcium ions flooding the sarcomere space.

On the other hand, we know that we can consciously contract or elongate our muscles; that is, we can reverse the above combinations. It is possible to voluntarily elongate or contract the muscle cell under the action of an increase in the outside force, and it is possible to elongate or contract the muscle as that outside force is decreased.

These changes are common occurrences when weight lifting. The triceps muscles elongate when "curling" a weight. As the weight reaches a certain position, the force acting on the triceps decreases and the muscle contracts. The biceps muscle undergoes just the opposite response: It contracts when the weight is curled, and elongates when the weight reaches a certain position, so that the force acting on it is smaller.

But how does an outside force cause such changes in muscle length? We know that muscles act in opposition, with the triceps elongating when the biceps contract and vice versa. Clearly some feedback of information is required, and here consciousness must enter the arena.

The Role of Consciousness Is Choice

Consciousness is perhaps the most basic concept in modern psychology, yet it eludes definition. One definition includes the special awareness of *proprioceptive* body signals, fantasies, and dream material. The greater the consciousness, the more ability there is to sense these signals.

The concept of proprioception was introduced by the British physiologist Charles Scott Sherrington early in the 20th century. Sherrington's research dealt with neurons—nerve cells capable of transmitting both sensory stimuli and motor impulses from one part of the body to another. Sherrington researched the functions of motor neurons, neurons which control the contraction and elongation of muscle cells. He was particularly concerned with the ways motor-neuron activity could be regulated by sensory feedback. Proprioception refers to sensory inputs arising in the course of centrally driven movements, when the stimuli to the receptors, attached to the muscle cells, are delivered by the organism itself—in other words, by conscious control. Sherrington chose the prefix "proprio-" from the Latin *proprius*, meaning "one's own." He believed that the major function of the proprioceptors was to provide feedback information of the organism's own movement.

Muscle proprioceptors are of two kinds. One type senses tension or forces acting on and within the muscle, the other senses elongation/contraction or the spatial extension of the muscle. The length receptors of muscles send fibers into the spinal cord to form synapses of motor neurons that terminate on the same muscles—a feedback loop. Thus, whenever the length of the muscle increases, a feedback takes place that gives rise to muscular contraction. Tension receptors actually sense force rather than elongation. Their action is inhibitory: When muscle tension increases, the receptors act on associated motor neurons to reduce it—the feedback loop tends to relieve the tension.

Thus, together, in the length and tension receptors we have a feedback loop that tends to reduce/increase tension and resist changes in the length of the muscle.

To see how this works, extend your arm at this moment straight out from your body parallel to the floor and hold it there. Because you must hold your arm against the pull of gravity, your arm will begin to get tired. Now involuntary (unconscious) decreases of tension will take place in the muscles opposing the pull of gravity (the arm slightly falls). This leads to an elongation of the muscle. The activity of the proprioceptive length receptors will now feed back information to the muscles to contract, while at the same time, because of the decrease in tension, the activity of the proprioceptive tension receptors will feed back messages to decrease activity. Thus, the two will now increase the tension in the muscle, and decrease or contract its length; the two receptors feed back synergistically, causing the muscle to contract under the action of an increase in tension.

But now the feedback works again. Involuntary (unconscious) increases of tension in the muscles are coupled with the muscle's contraction. The contraction will then deactivate the proprioceptive length receptors, and the increased tension causes an increase in the force-receptor inhibition that feeds back information to the muscles to decrease tension and elongate. Again, there is a synergy. The muscle relaxes by decreasing its tension and elongating. Without your conscious control your muscles and nerves will continue to oscillate messages causing the muscles to alternately contract and elongate, attempting to increase and then decrease muscular tension.

The role of feedback is extremely important in the actions of proprioception. It is also important in grasping how consciousness enters the motor-neuron muscle system. The fact that proprioception involves both force and elongation means that the system feeds back in parallel processes.

These are the consequences of unconscious or unwilled changes in the muscle-neuron system. In other words, you don't have to do something about them. The result of this is the regulated or sensible "feeling" that the muscle will automatically contract when the tension increases, and elongate when the tension decreases. The feedback controls the system.

However, this is not what takes place when you consciously change the muscle length as an outside force is applied to the muscle. When an outside force is applied and "you" consciously enter the feedback loop (you do work), two possibilities occur. The force receptor is pouring on the juice, continually attempting to increase force-receptor inhibition resulting in a decrease in muscular tension. The outside force, however, maintains the muscular tension in spite of the inhibition messages. In one case, the muscle contracts consciously when the force is applied, and contracts consciously as the force is maintained. In another case, the muscle contracts consciously when the force is applied, and elongates consciously when the force is maintained. In each case, equilibrium is reached, with the muscle remaining under stress or tension. This results in feedback, bringing the muscle to an equilibrium length under the action of the outside force. The muscle undergoes length changes with effort. The muscle is continuously under stress.

ENTER THE MIND

It is not obvious how the mind enters in voluntary control of the muscle. The difference between "mindless" involuntary movement and "mindful" voluntary movement is subtle. To grasp how this difference arises, we need to consider the neuron-muscle interaction as a problem in quantum physics, not classical physics.

With classical physics, the interaction is an all-or-nothing process: Either the neuron fires and a message is sent, or it is inhibited from firing and a message is not sent. With quantum physics, we add to the list the paradoxical possibility that the neuron fires and doesn't fire at the same time. Thus, in contrast:

Classical Response	Quantum Response
1. Fires	1. Fires
2. Inhibits	2. Inhibits
	3. Fires and inhibits

The third quantum response brings in the quantum nature of the interaction and provides the means by which consciousness enters the muscle.

The role of mind here is one of choice. With the involuntary case, no choice was needed: The actions are synergistic. The feedback loop is always a combination of two signals that act either to increase the force and decrease the length (an unambiguous signal to contract the muscle) or to decrease the force and increase the length (an unambiguous signal to elongate the muscle).

With the voluntary case where the muscle was elongated, the feedback loop was a combination of a force-decrease signal accompanied by a length-decrease signal. This is a double-contradictory message. These are the messages felt when, for example, a person is lifting weights, or, for that matter, willing his muscles to respond in a certain manner.

The willful action of elongating the muscle when the tension in the muscle has increased is opposite to the "unconscious" action that would tend to contract that muscle. Somehow the contradictory signal (which is the message to decrease tension and shorten length) is interpreted to contract the muscle while the tension is not decreasing. Thus, the muscle is given messages to both increase tension and decrease tension as well as to increase length and decrease length.

Schrödinger's Cat and More on the Observer Effect

Whenever a system contains two opposite messages, a choice must be made. That choice, I believe, resolves a quantum physical paradox familiar in quantum mechanics and known as the Schrödinger cat paradox.

The cat paradox arises when a living cat is placed in a box so that no one can see inside the box or influence the outcome of what happens in the box without opening the box. Inside the box is a radioactive element that either will or will not emit a signal in one hour. If it does, the cat will die; if it does not, the cat lives. This is because also inside the box there is a cyanide pellet poised above a beaker of hydrochloric acid, and if a signal is sent, the pellet drops; if not, it remains suspended above the beaker. Here is the paradox: After one hour will there be a live or a dead cat in the box?

Since radioactive decay is a quantum physical process, all one can say is that the radioactive element has a quantum wave function (qwiff) containing possibilities of both "decay" and "not decay." This, in turn, means that the pellet has a qwiff coupled to the element containing the possibilities "dropped in acid" and "not dropped in acid." And this couples in the cat's qwiff, "live" and "dead." As long as no one opens the box both possibilities exist! The cat is both alive and dead. As soon as the box is opened the cat is either alive or dead.

Many people find this "cat business" silly indeed. After all, doesn't the cat know if it is dead or alive? The point is that it does know it's alive in one of the possible worlds to be discovered by the observer, but in the other world, where it is dead, it probably doesn't know anything at all (unless there is a cat heaven or the like).

When the observer opens the box and finds a dead cat, the observer and the dead cat constitute one possible world. If, on the other hand, the observer finds a purring pussy, he and that cat are in a different world, parallel to the dead-cat world.

Although this is indeed strange, it is extremely difficult to think up any other alternative. Physicists have been debating the fate of the cat ever since Schrödinger devised the paradox in 1935.

Similarly, when a voluntary muscle is put under tension, both increase- and decrease-tension signals arise, resulting in *both* the potential firing and the inhibition of neurons. The neuron is in the familiar dual state of potentially having fired and not having fired at the same time. If no notice of the situation is taken, the increased outside force will "do its will" and the muscle will either contract or stretch depending on its location and without any awareness of the person. This could result in "unconscious" accidents.

For example, imagine yourself riding a bicycle for the first time. Your willful attempt to keep your bike up and running is in question as you pump your pedals faster. As your bike starts to tip, your muscles receive new information from your nerve cells. They are given signals both to contract and freeze before your fall and to expand, in spite of the consequences, letting your bike and you right yourself by simply turning into the fall. Without a conscious noting of what's going on, the muscles will probably contract and you will hit the dirt. But with your mind entering in, the muscle "decides" to elongate in spite of the signal telling it to contract. With further training, you "learn" not to contract those muscles when falling and just stay aboard, riding out the fall by turning the front handlebars.

However, by taking note of the contradictory signal, the paradox is resolved and the muscle contracts or it doesn't, and, simultaneously, we are *aware* of which event actually occurs. If the wrong action takes place, we note the wrong action. In this manner we begin to alter the probability patterns by will. Each time we correct the action until the desired result takes place.

Since we know that we can either contract or elongate our muscles once we are aware that they are under tension, taking note can have either action. Which action is taken probably depends on other feedback factors. In this manner we learn how to overcome outside forces, and in so doing we create our muscular signatures: our smiles, our frowns, our ups, our downs.

Thus, conscious control of muscles must involve quantum physics by invoking choice. Once choices are made that produce the desired action, the process becomes a habit and the conscious mind no longer "pays attention" to the process. Thus, in the first example, where an outside force was applied and the muscle elongated under stress, the conscious mind had to choose between the two contradictory messages. The result was the muscle contracted.

In the second example the quantum choice did not occur when the outside force was first applied. It occurred afterward but again produced a muscle contraction. The feedback loops are different because they are different conscious choices—different desires are manifested in the muscle cells. In the first case, the muscle is willed to elongate; with the second, it is willed to contract.

This helps explain how human will enters the body human. It is the first in-depth look at what I call the body quantum.

PART TWO

Eating

One memorable evening not too long ago, I found myself dining at one of my favorite Chinese restaurants in San Francisco with a number of dear friends. We hadn't seen each other for a long time, and were all eager to tell each other the latest news of our busy lives.

As the meal progressed, I began to realize that because of the distraction of our animated conversation, I was missing the enjoyment of the delicious food. At that moment I decided to tune out the conversation to dwell on the taste of my food. With each morsel, I tuned in and tried to identify just what was passing over my taste buds. My friends were a little perturbed by my sudden self-absorption. I just went on eating slowly and silently. Finally someone demanded that I share what was going on in my mind. I told them about my taste adventures, describing down to the molecular level how I was able to pick out the various seasonings, the texture of the succulent meats, the acid taste of the peppers, and even how the food "felt" as it entered my stomach. I even attempted to describe the feeling of my digestive juices as they bathed the food.

My friends were amazed. Apparently, they had never attempted to be that conscious of food. Soon they were all eating silently and, from the looks on their faces, with more contentment than ever before.

I realized then that eating is indeed a great pleasure—one that can be enhanced even further if you try to focus your awareness on the food. Tasting is a molecular process involving the quantum physics of molecular interactions. By learning more about what takes place at the level of molecules within our bodies, it is possible to gain a far greater enjoyment of even the simplest pleasures of life.

Whether we eat to live or live to eat, eating can be both an adventure and a necessity. If you think about it at all, aside from its associated pleasurable (and for some dieters, painful) consequences, eating is a mystery. Briefly put, eating is the transformation of nonliving substances into living substances and energy.

In this part we will look at eating both from the body's need for an external energy supply and how the material of foodstuffs becomes living cellular matter.

CHAPTER 7

You Are What You Eat

I, like many people, love to eat. Indeed, eating is a grand pleasure of my whole life experience. I can't imagine sticking to any long and drawn-out diet. Wine in moderate amounts, fine French and Chinese cuisine, and good old barbecued ribs all delight my palate. A meal is an experience of communion not only with my dining companions but with the food itself.

My description of what is to follow, the physics of food, particularly insofar as it makes its way past my lips and teeth, will not indicate the enjoyment of the many chemical and physical processes that take place during digestion, when food is transformed into energy and waste products. The idea that food might be "enjoyed" once it has passed the taste buds may seem odd. However, I believe that one of America's worst problems with regard to health and energy management in the body is caused by what I term "unconscious eating." Perhaps by learning more about how food is turned into energy, even the most abstract aspect of digestion and energy transformation can be enjoyed as a conscious experience. Maybe indigestion will turn out to be a mental phenomenon.

An Overview of Digestion

When food passes through the body, it undergoes a series of chemical or quantum physical transformations. Once it has been swallowed, it is no longer called food. It is called the *bolus*. The bolus, under the action of gravity and the wavelike peristaltic action of the narrow esophagus, is rhythmically pushed into the stomach. The stomach is capable of holding about 2 pints of food in a normal adult. When you overeat and feel stuffy, it is simply because you have stretched the stomach bag beyond its normal 2-pint capacity. The stretching of the stomach pushes on other organs such as the diaphragm and the heart wall. Since this can impede breathing and heart action, it is no wonder that the stuffy feeling occurs. It is a warning sign that life is threatened.

After leaving the stomach, the bolus enters the duodenum, the first part of the small intestine. It is approximately 10 inches long. Here, further processes of digestion go on. The bolus that leaves the duodenum is now called the *chyme*.

From this point on, all chymes look mostly the same regardless of what you have eaten. In the small intestine the major work of assimilation of the products of digestion occur. The small intestine is divided into three parts: the duodenum, the jejunum (about 8 feet long), and the ileum (about 12 feet long). Once the chyme has passed through the small intestine, it enters the large intestine. Here, in the normal well-fed adult, about 12 ounces of chyme enter per day. In the large intestine water is reabsorbed, leaving a mass called the feces, which is about two-thirds reduced in weight (about 4 ounces of feces). This estimate is quite variable; a diet rich in roughage would have considerably more fecal matter.

Why Do I Eat? The Energy of Food

What happens when I eat? Physics has shown that all activity—that is, all things that undergo changes in motion as a result of work—require energy. Before I get into energy, let me explain what "work" is. Work occurs whenever a force is applied through a distance. By multiplying the force times the distance over which the force acts, in the direction of the distance, one computes the amount of work accomplished. That work must be transformed into something. That something is called energy.

Energy appears in many guises. Every object in the universe has energy. This energy can be divided into two types: kinetic energy, and potential energy. Most objects have more potential energy than kinetic energy. This means that potential energy, in turn, appears in many forms. The largest potential energy contained in any physical object is its own mass. Recall Einstein's theory of relativity, $E = Mc^2$, where E is the energy, M is the mass, and c is the speed of light. When an object is at rest with respect to you, all of that E is called rest mass energy. When the object moves with respect to you, some of that E is kinetic energy and some is rest mass energy.

Actually this simplification is true only for objects that aren't moving too fast. In this case one says that E has two distinct parts, a rest mass energy, and a kinetic energy. If the object is composed of internal parts, such as a molecule, then the energy has three parts: a rest mass energy, a kinetic energy, and a quantum potential energy.

In the human body the rest energy usually does not change. Thus one need deal only with kinetic energy and quantum energy in normal digestion and in all other processes where energy transformation takes place.

The Thermodynamics of Eating

In order to grasp how food energy becomes body and cell energy, we need to look at some aspects of the two laws of thermodynamics:

Law I: Energy is neither created nor destroyed.

Law II: Energy tends to move from order to disorder.

THE FIRST LAW: YOU DON'T GET SOMETHING FOR NOTHING

The first law of thermodynamics is quite evident in the universe. It is nature's way of keeping books. It simply says that in all physical processes in the universe, transformation occurs when energy changes form. Thus, when a match burns, the energy contained in the hydrocarbon substance of the paper pulp of the match is released. The hydrocarbon combines with oxygen, forming carbon dioxide and water.

In the body, a common type of burning is the oxidation of glucose, which is a carbohydrate or sugar. Here, a single molecule of glucose interacts with 6 molecules of oxygen to produce 6 molecules of carbon dioxide and 6 of water. Of course, in actual practice within the body many molecules are involved. When it comes to such massive chemical reactions, a useful concept is that of the *mole*.

No, a mole is not a furry little creature but the weight in grams of 6×10^{23} molecules of the substance in question. This number is called Avagadro's number after Amadeo Avagadro, the Italian physicist who discovered it in the 19th century. When multiplied by the weight of a single atom of hydrogen, Avagadro's number equals 1 gram. There are about 454 grams in a single pound of weight. When hydrogen gas is formed, 2 atoms of hydrogen bond together. So 6×10^{23} molecules of hydrogen gas would weigh 2 grams. Following this line of reasoning, a mole of oxygen molecules would weigh 32 grams and a mole of glucose, 180 grams.

When a mole of glucose (180 grams, or 6.34 ounces) is burned with 6 moles of oxygen, it produces 6 moles each of carbon dioxide and water. It also gives off energy. In so doing, 686 calories are liberated. This energy is called the free energy, which means it is available to do useful work.

In a plant the reaction goes the other way: 6 molecules of carbon dioxide and 6 molecules of water produce a single molecule of glucose and 6 molecules of oxygen. Thus, plants synthesize molecules of glucose from carbon dioxide and water. Free energy must be added in order to make the reaction go. This energy comes from sunlight in the form of quantum units called photons. So actually, energy is conserved. Plants make glucose from sunlight, carbon dioxide, and water.

In a human body there are energy cycles. Energy flows from the energy-rich molecules in our foodstuffs to fuel the work of our cells. The force which drives these energy cycles is the second law of thermodynamics.

THE SECOND LAW: RUNNING THE UNIVERSE DOWN

Take an empty drawer. Carefully put your child's socks on one side of the drawer, leaving the other side empty. Close the drawer. Return one day later after your child has frantically searched through the sock drawer. What you find is socks in disarray on both sides of the drawer.

The socks always tend to be disturbed and distributed throughout the drawer after the child's interruption. The next day the child must spend more energy in just rearranging his drawer so he can find his socks now that he needs them.

Something is running down in order to get energy to go from one place to another place, and this running down into chaos has a measure. It is called *entropy*. The messy drawer is said to be at a higher entropy than the neat drawer.

The second law says all processes tend to go in the direction of increasing entropy. True, when the sun shines on a plant, the plant's entropy actually decreases. The plant becomes more orderly and structured in producing glucose. However, the sun has become more entropic. Its entropy increase outweighs the decrease in the plant's entropy. Like a spring mechanism, all energy processes wind down, producing more entropy.

When you eat, food becomes glucose, amino acids, and fats. Glucose is burned and changed into water and carbon dioxide, and in so doing, entropy increases.

Free Versus Expensive Energy

Previously I explained that energy came in two forms: potential, and kinetic. Both forms of energy are, however, not always available to do work. Sometimes energy becomes expensive because it is not readily available. For example, the heat energy contained in the ocean cannot be used to heat up our houses because our houses are usually kept at a higher temperature than the ocean. Work must be done on the ocean water in order to extract its energy and move it toward our homes. (In fact, this is why our refrigerators use electrical energy to cool food.) On the other hand, that same ocean energy will melt without added effort an occasional iceberg drifting southward from the North Pole. That same ocean energy unavailable to heat our homes is available to melt an iceberg. Thus, energy can be thought of as free or as expensive depending on what needs to be done to move it from one place to another. Consequently, there are not only two kinds of energy in the world there are really four, including energy able to do work and energy which, like a lazy worker, is not available for work at all. In a human cell, both pressure and temperature tend to remain constant or unchanging as physical processes occur. Thus, when a process takes place in which there is an energy transformation or energy change, there is a corresponding change in free energy, and another energy change which depends on the temperature of the cell and the change in entropy. This is called the expensive or entropic energy.

The total change equals the change in free energy plus the change in expensive energy. Entropic energy is expensive because it represents energy that has gone into processes that are not recoverable. When an ice cream cone melts on the pavement, most of the energy has gone into heat and evaporation, which is entropic energy, so little is available to do any work.

The Physics of Enzymes

We all know that a match will burn when struck. Certainly there is enough oxygen in the air and enough fuel in the match to enable the burning to occur. But why doesn't the match just spontaneously burst into flame? Obviously, burning a match releases energy. The reaction thus produces heat and water vapor, and therefore, according to the second law of thermodynamics, entropy is certainly produced. Thus, the reaction would appear to "go" without any need to strike the match in the first place.

The problem is not in the laws of thermodynamics, but in the chemical energy available for the burn. Just holding the match up in the air does not cause the match to leap into flame because there is an energy barrier to spontaneous combustion. To overcome this barrier, or any similar barrier to reaction, requires a means for lowering the barrier or finding enough energy to get over it. In the case of the match, the barrier is not actually lowered, but heat energy that is enough to overcome the barrier is added. But suppose as in the human body, there isn't enough heat energy to accomplish the task. In that case nature has designed molecules that accomplish the task. They are called enzymes.

Enzymes are required in nearly all biochemical reactions. They play the role of lowering energy barriers that exist between the initial reactants and the final products. Ultimately, the reaction leads to the production of energy.

An enzyme is a specialized protein molecule. Over 1,000 enzymes are known to exist in the human body. When an enzyme, E, interacts with an initial substance, S, called the *substrate*, it produces a product, P, in a two-step process:

$$E + S \rightarrow ES$$
$$ES \rightarrow E + P$$

In other words, the basic reaction, $S \rightarrow P$, is assisted by the enzyme E.

Each enzyme consists of a chain of amino acids, called a *polypeptide*, folded over into a complex globular structure. When the enzyme and the substrate combine, the enzyme is quantum mechanically deformed, producing a key-lock fit, ES. In so doing, the substrate molecule is also deformed, lowering the energy barrier. This barrier is called the *activation energy*.

Activation energy is commonly experienced when lighting a simple match. As is obvious, a match burns in oxygen and will continue to do so, producing heat, carbon dioxide, and water. If there was no activation energy present, the match would spontaneously burn in the presence of oxygen. But you must strike the match, providing the necessary heat energy to overcome the activation barrier.

WHY YOU NEED YOUR VITAMINS

An important enzyme in the body, also containing phosphorus, is vitamin B5, also known as pantothenic acid and as Coenzyme A (CoA). CoA acts enzymatically to break down long chains of fatty acids (fats) in order to make use of the products in the famous Krebs cycle of energy production. The basic action of the enzyme is to break apart long chains of fat molecules into shorter two-carbon links used in the Krebs cycle.

A long-chain molecule of fat, specifically palmitic acid, binds with CoA. Then the CoA gradually takes the fat apart piece by piece, forming units of aceytl CoA. One molecule of palmitic acid produces eight molecules of aceytl CoA. With each break in the chain valuable energy is released to fuel the body's needs.

When an enzyme comes into the vicinity of the substrate S, it binds with it temporarily and lowers the energy barrier. Usually this lowering is enough to allow the reaction to proceed. There is still some barrier present, but the small amount of heat normally present in the body will give enough of an energy boost to the ES partnership for the enzyme and substrate to burst apart. Only this time the substrate is transformed into the product, P, and lots of energy.

To make this more visible, think of the wood in the match and the oxygen in the air as the substrate. Think of the carbon dioxide, water vapor, and other products of burning as the product, P. The enzyme in this reaction is the striking surface, usually containing phosphorus.

Are you what you eat? Well, not quite. Not all of your food is turned into you—that is, the material you. Some of the food is converted into energy, which your material body uses to move.

The breakdown of foodstuffs usually requires enzymes. These enzymes, if not used up, are recycled so that they can be used again and again. Vitamins are important enzymes, and, thus, our bodies need them to assist in the processing and digestion of most of our foods. In each case the ultimate product of food breakdown is the production of glucose or blood sugar.

If we return again to Einstein's famous $E = Mc^2$, we can gain another insight into how energy is transformed in the body. The E stands for all energy, the M for all mass. Since this equation tells us that energy and mass are equivalent, it means that before energy is released, say by the oxidation of glucose, it is contained as mass in the glucose and oxygen. This extra mass is found in the energy-rich bonds of the molecules. After the oxidation, this extra mass appears as energy. Thus, in the reaction:

$$\text{sugar} + \text{oxygen} \rightarrow \text{carbon dioxide} + \text{water} + \text{energy}$$

the carbon dioxide and water actually contain less mass than the mass of sugar and oxygen.

This extra mass is found in the quantum physical bonds that hold the sugar molecules and oxygen molecules together. In the reaction, energy is released and entropy is created, resulting in the physical movement of life itself in the body human. It is another example of the body quantum at work.

CHAPTER 8

Energy and Time to Burn

We all become hungry after a while. Controlling our hunger is an invisible clock or rhythmic cycle. To grasp how we become hungry, let us look at the internal clocks ticking away inside us all.

We Got Rhythm

The key insight into our human time sense is rhythm. Rhythm is sensed as resonance with vibration or frequency. Something inside of us is vibrating at a fixed rate. As time passes, another rhythm is imposed, interacting with the steady rhythm of our life process. Whenever two different frequencies vibrate together, they produce momentary reinforcements called beats. These beats are felt as hunger, sleep, heartbeat, and other rhythmic processes. In other words, we resonate with vibratory processes—some of them occurring with slow repetition, others faster.

Consequently, each of us possesses a sense of time because, as Ira Gershwin might have stated it, "We got rhythm." Some of our rhythms include: (1) the sleep-awake cycle, (2) the body temperature cycle, (3) the oxygen-use cycle, (4) the heartbeat cycle, (5) the kidney-excretion cycle, and (6) the taste, smell, and hearing cycles.

The sleep-awake cycle is quite easy to observe. We all tend to rise each morning and fall asleep each night at about the same time. The amount of sleep needed appears to range from about 6 to 8 hours for most people.

The hypothalamus of your brain regulates a rhythmic variation of your body temperature of about 1.5 to 2 degrees, the lowest occurring during sleep (presumably at night) and the highest in the afternoon. People often find that during late-night TV watching they grow cold and require an extra blanket or covering. This is a signal that your temperature cycle is getting ready for the sleep-time shutdown.

Oxygen is needed by each and every cell of your body. Of

course, exercise and exertion increase that need and cause you to breathe harder. However, even when sitting still for a whole day, your body will show a rhythmic variation in oxygen consumption.

The most notable body rhythm is your heartbeat or pulse. In 1581, 17-year-old Galileo discovered, while standing in the Cathedral of Pisa, that a huge hanging chandelier made the same number of swings per heartbeat whether those swings had large or small arcs. He was checking the rhythm of the swings against his own heart as a standard. It is no coincidence that the human heart beats nearly at 1 beat per second when a person is at rest. Of course, the actual range of a normal resting heart is about 55 to 80 beats per minute. Drugs, such as aspirin, can also increase the normal rate. Probably, however, the origin of the second of time was the human heartbeat. Nowadays we use the clock to measure our heartbeats.

Kidneys filter the blood passing through them and remove waste products, which are eliminated as urine. Although it is not unusual to have to rise during the night to urinate, it is rare that this function needs to be repeated. Most often, the kidney performs at a higher rate during the day and slows down at night.

The senses of taste, smell, and hearing are usually dull in the morning but sharp by the late afternoon or evening. That is why we tend to prefer a bland breakfast and more flavorful lunches and dinners.

These and other body rhythms give us our human sense of time. Without patterns or rhythms we couldn't have a time sense.

What determines our body rhythms? In a research experiment, 130 human beings lived together in caves without the light of day and without clocks or other timekeeping instruments. They all adopted a 25-hour day as a natural rhythm. (Burns, p. 105.) Why this is 1 hour greater than the earth's rotational period is a mystery. Perhaps the answer is that we aren't from this planet but from another one where the daily cycle is 25 hours long!

Metabolism and the Rate to Life

Similarly, the rhythm of life enters into metabolism—the rate at which we convert food into energy. We each burn the stored or potential energy of food at a certain rate—so many calories in an hour or in a day. This is how we convert stored energy into kinetic energy—the energy of body movement and heat. This energy is measured in calories. The calorie I'm discussing is really a kilogram calorie, the amount of heat needed to raise the temperature of a kilogram of water by 1 degree centigrade. A calorie is a unit of energy equal to 4,184 joules. One hundred joules is the amount of energy burned up by a 100-watt light bulb in just 1 second. Burning 92 calories per hour—which is exactly what a 180-pound human being consumes at rest—is equivalent to the output of a 107-watt light bulb in that same hour.

The lowest rate of energy consumption is called the *basal metabolic rate*, or BMR. It is the amount of energy needed to perform minimal body functions such as breathing, pumping of the heart and blood through the blood vessels, seeing, and relatively small movements of the body.

Once this energy is burned, setting or keeping in motion whatever is required, a byproduct is produced: heat. It is a consequence of the second law of thermodynamics that whenever there is an energy conversion resulting in work, heat must also be produced. It is the surface area of the skin (around 20 square feet) that acts as the heat-radiation device. The larger the surface area, the greater the heat radiated per pound of weight. Consequently, the size of an animal affects the rate at which energy can be consumed. Research has shown (Cameron, p. 92) that as the body weight of an animal increases, the BMR also increases. (See Table 1.)

Thus, the larger the animal, the greater is its BMR. A person has a BMR of about 2,000 calories per day. A cow, which weighs about five times as much as a person, has a BMR of 6,700 calories per day, while a tiny rat weighing half a pound has a BMR of only 30 calories per day. Friendly Fido weighing in at 40 pounds burns 720 calories per day.

Thus, there is an advantage to increasing the size of an animal. By reducing the surface-area-to-weight ratio, the radiated heat is reduced. Fewer and fewer calories are needed for warmth.

Table 1: BMR vs. Animal Size

Animal	WT	BMR*	BMR/WT
Tiny Rat	.5 lbs.	30	60
Friendly Fido	40 lbs.	720	18
Human Being	200 lbs.	2,000	10
Cow	1,000 lbs.	6,700	6.7

*BMR (calories per day)

Body Heat and Thermodynamics

All dynamic processes are associated with a variation of body energy in some form or another. Yet if we look carefully at each process, we find that as time passes, the total energy involved is conserved or maintained overall. As time marches on, energy transforms from one form to another. The most important relationship governing the change of energy is the first law of thermodynamics:

decrease in energy = heat loss + work done

This equation must be obeyed for each system of the body, as defined by spatial boundaries. In other words, if a region of space within the body has a defined boundary, it is a system. A cell is a system. An organ is a system. And of course the whole body is a system.

In general, the work done by a small system makes energy changes for the larger system in which it exists. Energy changes of that system appear either as heat or as external work performed by the yet larger system in which it exists. For example, a cell performs the work of changing glucose into carbon dioxide and water. To do this work, it takes in heat from outside cell tissue and produces the kinetic energy of the molecules within the cell.

Suppose that the cell is a heart cell. The work performed by the heart cell actually turns out to be quite small in comparison with the decrease in heart cell energy per second. To pump blood, the whole heart produces .0143 calorie per minute, or 238 microcalories (a millionth of a calorie) per second. To do so, it must undergo an energy transformation of .08 calorie per minute, or 1.333 microcalories per second, nearly six times as much energy change to accomplish the work. Where does the rest of the energy go? It goes into heat. It is a general observation that most of the energy changes in a biological system such as an organ of the body are radiated away as heat.

The total energy changes of all the organs comprise the basal metabolic rate. The BMR measures the base rate at which energy is being transformed into heat and the work of the internal organs when the body is at rest. The BMR is divided into various functions, such as the work of beating the heart and using the skeletal muscles (25 percent), making the kidneys function (10 percent), thinking and other brain activity (19 percent), and keeping the spleen and liver working (27 percent). The remaining organs use up 19 percent of the energy available to do work.

Another way to determine the BMR depends on oxygen consumption. Experiments show that as the rate of oxygen consumption increases, so does the BMR. Oxygen consumption is measured in terms of volume per minute per pound of body weight. A sleeping human uses about 1.43 milliliters (ml, a thousandth of a liter) of oxygen per pound of weight each minute. Each of our organs also uses oxygen. Table 2 shows just how much oxygen, energy, and percentage of BMR is used by a resting human being weighing 143 pounds.

Of course, each of us accomplishes more in life than just keeping our organs functioning. If we examine some standard activities (Table 3), we find that the metabolic rate increases depending on the activity. The exception to the rule is sleeping, which actually burns less than the BMR. If we take into account several possible activities, we can devise an activity analysis of a typical person depending on body weight.

To use Table 3 for yourself, all you have to do is multiply the activity entry in the table by your own body weight. For example, suppose you weigh 200 pounds. Then your sleeping oxygen consumption is 1.43 × 200 = 286 ml/min. If you weigh 150 pounds, this number would be only 214 ml/min.

In 8 hours of sleeping you (weighing 200 pounds) would need to breathe in 137 liters of oxygen. Since oxygen comprises 20 percent of the air taken in, this means you would breathe in 685 liters of

Table 2

Human Organ	Oxygen Used (ml/min)	Energy Used (cal/min)	BMR %
Brain	47	.023	19
Heart	17	.08	7
Muscles	45	.22	18
Liver	60	.30	20
Spleen	7	.03	7
Kidney	26	.13	10
Others	48	.23	19
	250	1.22	100%

(Adapted from R. Passmore in R. Passmore and J. S. Robson, eds., *A Companion to Medical Studies*, Vol. 1. England: Blackwell, Osney, Mead, 1968, p. 49.)

Table 3

Activity	Oxygen	Energy	Energy
	Used (ml/min/lb) (1,000 ml = 1 liter)	Used (mcal/min/lb) (1,000 mcal = 1 cal)	Used (cal/day/lb)
BMR	1.75	8.53	12.28
Sleeping	1.43	7.18	10.34
Sitting at rest	2.03	10.2	14.69
Standing relaxed	2.15	10.8	15.55
Sitting alert	3.59	18	25.92
Walking slow (3 mph)	4.54	22.7	32.69
Golf	5.28	26.5	38.16
Bicycling (10 mph)	6.82	34	48.96
Tennis	7.53	37.7	54.29
Swimming (1 mph) Crawl	6.97	35	50.4
Breaststroke	8.11	40.7	58.61
Backstroke	8.76	44	63.36
Climbing stairs (116 steps/min)	11.7	58.6	84.38
Bicycling (13 mph)	12	59.8	86.11
Running (5.7 mph)	12.7	63.7	91.73
Running (7 mph)	15.3	77	110.88
Running (11.4 mph)	22.9	115	165.60
Swimming (2.2 mph) Crawl	28.3	142	204.48
Breaststroke	32.5	163	234.72
Backstroke	35.2	177	254.88

(Based on P. Webb in J. F. Parker and V. R. West, eds., *Bioastronautics Data Book.* Washington, D.C.: National Aeronautics and Space Administration, 1973, pp. 859–61. Also on Laurence E. Morehouse, Ph.D., and Leonard Gross, *Total Fitness in 30 Minutes a Week.* New York: Simon & Schuster, 1975, p. 182.)

air. During your 8-hour sleep, you would burn 689 calories (7.18 millicalories per minute per pound of weight × 200 pounds × 60 minutes × 8 hours).

Suppose that now you got up to watch television for 8 hours. You would increase your oxygen consumption by breathing in 1,723 liters of air. You would also burn up 979 calories (10.2 millicalories per minute per pound of weight × 200 pounds × 60 minutes × 8 hours). So far, in 16 hours you have used up 1,668 calories.

During the next 8 hours of your day, let us assume that you walk for 1 hour (272 calories = 22.7 millicalories per minute per pound of weight × 200 pounds × 60 minutes), stand for 3 hours (389 calories = 10.8 millicalories per minute per pound of weight × 200 pounds × 180 minutes), and sit at your desk doing a normal amount of light work for 4 hours (864 calories = 18 millicalories per minute per pound of weight × 200 pounds × 240 minutes). Your total caloric expenditure has gone up to 3,193 calories. If you weigh only 150 pounds, however, you would have consumed only 2,395 calories in doing these same activities.

As an added convenience in using Table 3, I have included a column showing the energy burned per day per pound of body weight. Thus, the BMR of 8.53 millicalories per minute per pound translates into 12.28 calories per day per pound. To determine your BMR, simply multiply 12.28 by your weight in pounds.

In Table 4, I have taken the same activities and listed the number of calories burned per minute depending on your particular body weight. The table's entries are shown in 10-pound intervals, starting with a 100-pound person and ending with a 220-pound person. Thus, for example, suppose you weigh 150 pounds. The number of calories you burn per minute doing nothing is 1.28. If you decide to go for a walk to the corner drugstore at a leisurely pace of 3 mph, you will burn up 3.40 calories per minute. If your walk takes 10 minutes, that comes to 34 calories burned. Suppose next that you stand in the drug store reading magazines. Assuming that the magazine you pick up isn't too stimulating, you will burn up 1.62 calories each minute you stand there. If the magazine really gets you hot, you will probably burn a little more.

By taking into account all of your activities, it won't take too long to figure out your calorie day. Just multiply the number of minutes you perform any particular activity listed and add up to get the result. The number might surprise you. If that number exceeds your caloric intake, you will lose weight. Your body will need to dig into its store of fat to provide you with the energy for the day.

If I take myself as an example, I can determine my calorie day. Suppose I decide to spend 24 hours in a comatose or deeply mediative state. I weigh 165 pounds. Accordingly, I multiply the BMR energy used, 12.28 cal/day/lb, by 165 to determine my BMR expenditure of energy. This comes to 2,027 calories per day. Suppose I decide to eat 2,400 calories and neglect the work performed in digesting the food, the increase in body temperature associated with this diet, and other efforts I might perform during that day. Then each day I would have a surplus of around 400 calories to deal with. That surplus becomes excess body weight. In the next table we see just how much excess is involved.

Table 5 shows just how dense in calories foods are. A pound of carbohydrates contains 1,864 calories, a pound of fat 4,227 calories, while a pound of protein contains 2,409 calories. By eating an excess of 400 calories each day, my body weight would increase. Just how much that increase would be depends on what happens to those calories. If all of them became fat, then it would take me about 10 days to gain 1 pound. In 1 month my weight would jump to 168 pounds, and in 1 year my weight would have climbed to nearly 200 pounds. (Does this sound familiar?)

As you can see, it doesn't take much to gain weight if you eat more than you burn. This calculation, however, isn't quite accurate. Since my weight is increasing each day, my BMR is also increasing each day. The more I weigh, the faster I burn. Thus, with each pound of weight gained there is a compensating factor which increases my burning rate.

That is why people who are extremely overweight find it easy to shed large amounts of weight in the initial stages of dieting. For example, a 350-pound person burns with a BMR of 4,300 calories per day. By eating a healthy diet of 2,400 calories per day, this person would in principle shed nearly half a pound a day of fat (assuming that the person sheds fat and not protein). Thus, in 1 month his weight would drop by 15 pounds of fat; in 1 year, he would drop 180 pounds of fat. His actual weight loss could be much higher considering the drop in water weight as well as fat. Or, it could even be considerably more if the person sheds protein as well as fat, since protein contains fewer calories per pound. (Thus, it takes more pounds of proteins to shed the same number of calories.)

It must be noted, however, that this calculation is not entirely accurate. To determine more accurately a BMR/fat drop/weight gain profile, we need to consider both the change in BMR and the calorie drop or gain per day. By taking into account the change in the BMR as the person's weight changes, I find that the 165-pound

Table 4: Calories Burned in 1 Minute of Activity

Activity	Weight												
	100	110	120	130	140	150	160	170	180	190	200	210	220
BMR	.853	.938	1.02	1.11	1.19	1.28	1.36	1.45	1.53	1.62	1.70	1.79	1.88
Sleeping	.718	.790	.862	.933	1.00	1.08	1.15	1.22	1.29	1.36	1.44	1.51	1.58
Sitting at rest	1.02	1.12	1.22	1.33	1.42	1.53	1.63	1.73	1.84	1.94	2.04	2.14	2.24
Standing relaxed	1.08	1.19	1.30	1.40	1.51	1.62	1.73	1.84	1.94	2.05	2.16	2.27	2.38
Sitting alert	1.80	1.98	2.16	2.34	2.52	2.70	2.89	3.06	3.24	3.42	3.60	3.78	3.96
Walking slow (3 mph)	2.27	2.50	2.72	2.95	3.17	3.40	3.63	3.86	4.09	4.31	4.54	4.77	5.00
Golf	2.65	2.92	3.18	3.45	3.71	3.98	4.24	4.51	4.77	5.04	5.30	5.57	5.83
Bicycling (10 mph)	3.40	3.74	4.08	4.42	4.76	5.10	5.44	5.78	6.12	6.46	6.80	7.14	7.48
Tennis	3.77	4.15	4.52	4.90	5.28	5.66	6.03	6.41	6.79	7.16	7.54	7.92	8.29
Swimming (1 mph) Crawl	3.50	3.85	4.20	4.55	4.90	5.25	5.60	5.95	6.30	6.65	7.00	7.35	7.70

Breaststroke	4.07	4.48	4.88	5.29	5.70	6.11	6.51	6.92	7.33	7.73	8.14	8.55	8.95
Backstroke	4.40	4.84	5.28	5.72	6.16	6.60	7.04	7.48	7.92	8.36	8.80	9.24	9.68
Climbing stairs (116 steps/min)	5.86	6.45	7.03	7.62	8.20	8.79	9.38	9.96	10.55	11.13	11.72	12.31	12.89
Bicycling (13 mph)	5.98	6.58	7.18	7.77	8.37	8.97	9.57	10.17	10.76	11.36	11.96	12.56	13.16
Running (5.7 mph)	6.37	7.01	7.64	8.28	8.92	9.56	10.19	10.83	11.47	12.10	12.74	13.38	14.01
Running (7 mph)	7.70	8.47	9.24	10.01	10.78	11.55	12.32	13.09	13.86	14.63	15.40	16.17	16.94
Running (11.4 mph)	11.5	12.65	13.80	14.95	16.10	17.25	18.40	19.55	20.70	21.85	23.00	24.15	25.30
Swimming (2.2 mph) Crawl	14.2	15.62	17.04	18.46	19.88	21.30	22.72	24.14	25.56	26.98	28.40	29.82	31.24
Breaststroke	16.3	17.93	19.56	21.19	22.82	24.45	26.08	27.71	29.34	30.97	32.60	34.23	35.86
Backstroke	17.7	19.47	21.24	23.01	24.78	26.55	28.32	30.09	31.86	33.63	35.40	37.17	38.94

Table 5

Food	Calories per Pound
Carbohydrates	1,864
Proteins	2,409
Fats	4,227

man's weight increases to only 167 pounds in 30 days, and reaches only 185 pounds in a year. Maintaining this diet forever, he would actually reach a passive target or equilibrium weight of 195 pounds after many years.

The 350-pound individual would lose weight on the 2,400-calorie-per-day input. His weight would drop to 337 pounds after 1 month, and he would weigh 249 pounds after 1 year of this diet. Again, because his caloric intake is the same as mine, his passive target weight would also be the same 195 pounds—a weight he would reach after many years.

But in making these assertions, I have assumed a rather comatose individual whose only life processes are those composing the BMR for burning calories. As we have seen, many activities burn up many more calories than the BMR. Just sitting alertly or working at your desk burns up nearly 26 calories per day per pound, as compared to the BMR of about 12 calories per day per pound. Depending on your personal activity profile, you could either gain or lose weight for a given caloric intake. The increase in activity results in a different target weight called the active target weight. Increased activity always results in a decrease in the active target weight.

Another important aspect to weight and diet control is the age of the individual. The older we get, the lower the BMR. A 2-year-old child has a BMR of about 23 calories per pound per day, while a 70-year-old person has a BMR of only 10 calories per pound per day.

As an added feature in this book I have included in an appendix a small computer program written in BASIC that will enable you to obtain an approximately accurate activity profile showing you what will happen to your own weight according to the number of days you diet and the number of calories you take in per day. Also included are the passive and active target weights. A typical output from this program is shown on the page opposite.

An activity/passivity weight profile for FRED
My BMR is 11.1538 and my age is 50
My present weight is 165
My caloric intake per day is 2,400
The number of days I want to consider is 60
My passive target weight in this diet is 215.172
My weight will be 172.347 after 60 days of passive dieting.
My activity list follows
How many hours per day SLEEPING 8
How many hours per day SITTING PASSIVELY AS IN WATCHING TV 4
How many hours per day SITTING ACTIVELY AS IN DESK WORK 6
How many hours per day STANDING RELAXED 1
How many hours per week WALKING AT 3 MPH LEISURELY 2
How many hours per week BICYCLING AT 10 MPH 0
How many hours per week RUNNING AT 5.7 MPH 1
My active metabolic rate, or AMR, is 16.1025
My active target weight in this diet is 149.046
My weight will be 161.74 after 60 days of active dieting.

An activity/passivity weight profile for JUDY
My BMR is 11.5 and my age is 44
My present weight is 149
My caloric intake per day is 2,000
The number of days I want to consider is 60
My passive target weight in this diet is 173.913
My weight will be 152.752 after 60 days of passive dieting.
My activity list follows
How many hours per day SLEEPING 8
How many hours per day SITTING PASSIVELY AS IN WATCHING TV 4
How many hours per day SITTING ACTIVELY AS IN DESK WORK 3
How many hours per day STANDING RELAXED 2
How many hours per week WALKING AT 3 MPH LEISURELY 10
How many hours per week BICYCLING AT 10 MPH 0
How many hours per week RUNNING AT 5.7 MPH 0
My active metabolic rate, or AMR, is 15.0563
My active target weight in this diet is 132.835
My weight will be 145.889 after 60 days of active dieting.

The results of this program are based on some quite simple assumptions (as physicists are prone to make). For example, I have assumed in Fred's activity profile that any weight change for him is due to fat only. Thus, Fred's weight loss or gain is due to fat and fat alone. Of course, on any diet both fat and protein as well as water are lost or gained. Also I have assumed that the BMR is dependent on total body weight. However, as a little thought will make clear, fat cells in the body may not carry out any metabolism—they are simply storage cells. Thus, a truer BMR may be based on the lean body weight, a weight typically about 75 percent of the total body weight.

Of course, the lean body weight may be more or less than 75 percent of the total, depending on several factors. Women typically have a lower lean body weight than men; that is, a man and woman each weighing the same would accordingly have different lean body weights. The man would have a higher lean body weight than the woman.

Naturally, many other factors influence the manner in which body weight changes according to diet and activities. For example, during the first half hour of aerobic exercise, 67 percent of our body's fuel needs come from fat metabolism, while 33 percent come from aerobic glycolysis (the conversion of blood sugar into energy). After a half hour this ratio changes to 50 percent fat burn and 50 percent sugar burn. As we exercise more, the glycolysis starts to reduce and we begin to burn more fat once again.

Thus it is that long-distance runners are never fat. Since they run for more than a half hour's duration at one time, they are continually reducing their body fat, converting it to the energy they need. It is also true that long-distance runners usually need high carbohydrate intakes just before a race, probably because they have few fat reserves in comparison with non-long-distance runners.

For ordinary people, those that don't wish to exercise aerobically for more than a half hour at a crack, there is hope! Since 67 percent of the body's energy needs are met by fat metabolism during the first half hour, exercising for just that period will reduce body fat and not significantly lower your blood sugar. Thus, you can exercise up to a half hour and not necessarily become overly hungry.

Some physicians do not believe that our body fat has anything to do with the amounts of food we eat. Known as the set-point theory, many believe that our bodies choose a set point for muscle-fat density, and it is probably set by our genes. In other words, our bodies take on the weights they do to match this muscle-fat density. For some people this ratio is 75 percent muscle and 25 percent fat, for others it is different.

Factors such as genetic disposition cannot be discounted. My approach to body-weight management has been based simply on the laws of energy conservation. Thus, without a change in your activity profile, there is little you can do to change your weight permanently. Following the set-point theory, just reducing your caloric intake will not be sufficient because you will probably correspondingly decrease your activity, thus lowering your metabolic rate. By increasing your activity, you can increase metabolic rate. The laws of physics come to your aid, and by maintaining your caloric intake, you will lower your weight if you become more active. Increased activity will undoubtedly make you feel better about yourself. If your present muscle-fat density ratio tends to a higher fat percentage than you presently enjoy, increased activity may lower your set point.

But even if it doesn't, increased activity will increase your metabolism. So if you are dieting, I would suggest at least a full half hour of some high-intensity exercise each day. A simple brisk walk will suffice to beat the set point, which tends to lower your metabolism.

Don't worry, the laws of physics are on your side.

PART THREE

Building

Not all of our food is turned into energy, carbon dioxide, water, and waste products. Some of it remains in the body—indeed, as our mothers told us as we went off to school on a cold morning, "You need to eat a good lunch so that some of it will stick to your ribs." Some of it really does stick to our ribs, providing muscle tissue, bone, and fat for insulation and energy storage, just in case we miss a meal.

Here is where the body exhibits a remarkable intelligence. How much of our food should go into energy and how much should be used to make our physical bodies? To understand these questions, it is necessary to grasp the quantum processes of building the body beautiful. Here, entropy is reduced as the elements of life are reorganized. To accomplish this feat, the body must do work. This is similar to the entropy reduction that occurs in a refrigerator. Ordinarily, heat energy moves from a hotter body to a colder one. For example, a cube of ice melts; in so doing, the organized crystals of ice heat up as the air warms the cube. The entropy steadily increases as the ice becomes water. In the fridge, just the opposite occurs: Warm water is turned into ice crystals. But to accomplish the task, the motor in the refrigerator must do work, sucking out the entropy from the water and releasing it to the outside air. The water lowers its entropy, but the rest of the universe pays the price.

Similarly, when our bodies are building, entropy is being

sucked out, and organized intelligent structures of living carbon compounds are created. Again, food energy is used as the fuel to fight the entropy battle.

The most basic unit of the body is protein, and proteins are made inside our cells. In fact, a good way to picture a cell is to imagine it as a factory manufacturing protein. Protein molecules look like long chains—you can think of them as sticky strings that can ball up, much as the strings and ribbons we save from gift packages. The stickiness of the protein strings is quite useful, enabling them to bind together into complex patterns of solid substance, much the way strings form a base for birds' nests.

Proteins are made of even smaller units called amino acids. Amino acids can link together into long chains: the molecules of life. These molecules are left-handed, acidic, and tied to nitrogen atoms. Like shy bathing beauties, some amino acids are water-fearing, some water-loving, while others are indifferent to its presence. Each life-acid unit has its own particular structure, and from just 20 of these units all human protein is built.

To build a protein requires information; that information is contained in a "quantum library" within the cell. To read one of its "books," a quantum mechanical reading machine is needed. DNA and RNA contain both the codes of life and the means by which their messages are read.

As John Donne once put it, "I am a little world made cunningly of elements . . ." And we are made of elements—atoms which are designed uniquely to perform geometric and electrical tricks. No trick is better than the dance of electrons within these atoms; indeed, it is this dance that makes better things through chemistry. All molecules follow electron laws of shells and bonds—the expression of electron hatred for itself and others like it.

And underlying all this is consciousness, and the body quantum, entering through the observer effect. In this matter, each cell "knows" about the other cells; in turn, each atom in each cell "knows" about the other atoms. It is, I believe, in this knowledge, this ability to alter the odds by choice, that all life is possible. And it is possible for the body to build itself—to alter its building "codes" to correspond to

70

the many active and conscious choices that we bring to our everyday existence.

In Part Three I will explore how consciousness, through the body quantum, enters into the building of the human body.

CHAPTER 9

The Quantum Physics of Building Up the Body

You Are What You Eat and What You Think

We know that we need food to live. Food provides the energy we need simply to live—to move the body and to carry out the requirements of kinetic life. And movement occurs not only at the level of arms, legs, and muscles, but also at the human cell level. To simply move molecules around takes energy.

But energy is also required for something else. As long as food is converted into energy through the buildup of ATP, and, in so doing, complex carbohydrates, proteins, and fats are broken down into simpler two-carbon units to be used by coenzyme A, entropy increases in the body. Organized structures are broken down in order to free up energy.

To build up the body, that is, to make products such as proteins, nucleic acids (DNA and RNA), polysaccharides (needed for making glycogen), and lipids (the storage fats found in adipose tissue around your waistline), entropy must be overcome. Doing so requires energy, just as running a refrigerator to send heat from a cold body out into the warmer air requires energy.

The hierarchy runs from basic building blocks to cells. Each step along the hierarchy is one of greater and greater complexity, and to grasp the process is impossible without assuming that some form of consciousness acts at each level of the hierarchy. The reason is obvious: Systems that reverse entropy can do so only at the expense of information. That information must be supplied from the outset.

At the bottom step of the hierarchy, carbon dioxide, water, and ammonia build into amino acids, nucleotides, monosaccharides, and fatty acids. It is not enough that such a step is possible. There must be a purpose behind it. That purpose may be explained by a principle known in quantum physics as *quantum correlation*. A quantum correlation arises whenever two separate units, such as atoms, molecules, or subatomic particles, interact with each other. Before their interaction, the two units are said to be uncorrelated. Whatever happens to one of them is quite independent of what happens to the other. But after they interact, they enter into a correlated state. This state contains more information about the two units acting together. This information is actually greater in a numerical sense than the total information contained in both units before they interacted. For example, when two hydrogen atoms interact and give off energy, they form a hydrogen molecule; this molecule can now do something that the two separate atoms cannot. Together, the two can vibrate, each moving toward and away from each other with respect to their common center of gravity; together, they possess vibrational energy. Separated, they have only kinetic energy moving independently of each other. The new vibrational state turns out to have less entropy and, therefore, more information than the two taken separately. Quantum physics indicates that simpler molecules build into complex molecules by quantum correlations that tend to reduce entropy and store information in the form of patterns of complex interparticle motions, vibrations, and electron-photon interactions.

Next, these correlated smaller molecules build into correlated macromolecules, such as DNA, RNA, proteins, polysaccharides, and lipids. And these build into supramolecular, highly quantum-correlated systems consisting of membranes, enzyme systems, and ribosomes. At this level, life begins to appear and make sense of itself. From the supramolecular life arise classically committed structures able to interact through material means. These constitute organelles made up of cell nuclei, mitochondria, and endoplasmic reticula.

When the cell emerges from all this, it already has a rather complex form of consciousness. It knows quite a lot about itself, and yet, like you and me, it doesn't know what it is made of. From here cells build organs and organs build human beings.

How does a cell know itself? The evidence lies in its ability to respond specifically to its environment. Each of the 60 trillion cells that make up our bodies contains the same genetic code. Yet each type of cell plays a specific role. The heart muscle cell performs differently from the stomach muscle cell. Even in a fetus six weeks from conception, heart cells begin to beat; this beating synchronizes as soon as there are two heart cells. Through biofeedback we know that self-knowledge modifies self. At every level of existence there appears to be a self-recognition pattern that modifies the life unit. Cells are altered by other cells' presence. The

73

alteration of cellular behavior cannot be due just to the presence of DNA.

Think of DNA as a piano: All the keys are present, but which vibrational patterns should be played? How should the quantum wave functions making up the code of life be altered? The answer seems to lie in conscious actions which dictate how a cell is to metabolize and how it is to build and reproduce itself. Nothing exists in a vacuum. The cell is aware of itself in the bath of other cells like itself.

The Bookkeeping Quantum Dynamics of Body Building

Each second of life, energy is utilized. A major unit in moving energy around the body is the molecule known as ATP, adenosine triphosphate. Consisting of three phosphate molecules, it acts like a miniature train carrying potential energy as its cargo. Every time a phosphate "cargo car" is unloosed, energy is given off. It is useful to consider cellular construction using ATP as the energy-supplying mechanism.

ATP is used not only to provide the energy we need to move our bodies, conduct nervous electrical energy, and keep our brains functioning, among other energy-taking processes, but also to build up our bodies. It is always easier to tear down the body and burn up energy in doing so than to build it up. Tearing it down releases bound energy from correlated states and frees it, producing a kinetic chaos. Building it up takes in energy, even though when some structures are built they may give off energy. The reason has to do with the second law of thermodynamics: It takes work to organize anything. The movement and transformation of energy in biological systems is called *bioenergetics*. When that energy is used for building up biological systems, it is called *biosynthesis*.

Lehninger, in his book *Bioenergetics*, has provided an inside look at the time development of biosynthesis. He looks at the dynamics of the cell *Escherichia Coli*. Known simply as E. Coli, it is a bacterium commonly found in the intestinal tract. Many of the processes going on in our cells are similar to those occurring in E. Coli. In what follows, I want to show you just how E. Coli does the job of building itself using ATP as its energy source. The figures that follow are simply the energetic bookkeeping for the cell; if you wish, skip the details and go to the summary tables just ahead. My point is to show that the majority of energy used by E. Coli actually goes into building itself through the construction of protein. Larger molecular units are used to keep track of everything. DNA and RNA molecules must also be created every time an E. Coli cell splits in two—undergoes mitosis.

In order to accomplish this, the little cell cannot be just a machine following orders; it needs to adjust and keep track of exactly what it is doing at all times. It requires a cellular consciousness—a kind of cell-mind of its own. I believe that this mind could well be the result of the quantum observer effect, causing both the correlation and the destruction of the correlation—"consciousness" observing, even at this minute level of living organization.

SOME QUANTUM CELLULAR STATISTICS

The simple E. Coli cell is about 1 micron in breadth and width, and 3 microns long. (A micron is 1 millionth of 1 meter, or approximately 39 millionths of 1 inch.) When alive and filled with water, it weighs about 10^{-12} grams (a gram is about .03 ounce); dried out, it weighs about one-fourth as much. Its chief ingredients of life are DNA, RNA, protein, lipids, and polysaccharides (sugar molecules connected together). Every 20 minutes (or 1,200 seconds), E. Coli undergoes mitosis, splitting into 2 copies of itself, a process requiring that every ingredient in the original be duplicated.

Dry E. Coli has a molecular weight, M, of about 1.5×10^{11}. This weight gives the number of equivalent protons contained within it, which means that if all the molecules in E. Coli were broken down into hydrogen atoms (protons with electrons, which weigh very little in comparison), there would be 1.5×10^{11} protons.

A single molecule of DNA, however, has a molecular weight of about 2 billion equivalent protons, a molecule of RNA, about 1 million. Continuing in this vein, a molecule of protein weighs around 60,000, a lipid molecule, 1,000, and a polysaccharide sugar, 200,000.

The above molecular weights for E. Coli are only approximations. There also are several forms of RNA molecules and DNA molecules, and many varieties of proteins as well. These numbers represent averages or estimates at best, holding somewhat true for all human cells.

Biological assays of dry E. Coli show that it consists of approximately 5 percent DNA, 10 percent RNA, 70 percent protein, 10 percent lipids, and 5 percent polysaccharides. Given the molecular weights of each present constituent, m, and the above percentages, pw, it is quite easy to arrive at the number of molecules, n, of that ingredient in the cell. (The formula is n = pw × M/m, where n is the number of molecules of a particular component, pw the percentage of that component present, M the E. Coli

molecular weight, and m the molecular weight of the component.)
Table 6 shows the results:

Table 6

Component	Percentage (pw)	Molecular Weight (m)	Number of Molecules (n)
DNA	5	2,000,000,000	4
RNA	10	1,000,000	15,000
Protein	70	60,000	1,750,000
Lipid	10	1,000	15,000,000
Sugars	5	200,000	37,500

When the cell is wet it also contains around 25 billion water molecules, each with a molecular weight of 18.

For the cell to be able to reproduce itself, the numbers under the column n must be doubled (it takes E. Coli the aforementioned 20 minutes to accomplish the job). Lehninger has estimated the number of ATPs, or n-ATP, needed to make a single molecular constituent. For example, to make a phospholipid with a molecular weight of around 1,000 takes the energy of about 7 ATPs, while a polysaccharide requires about 2,000 ATPs. A single molecule of DNA needs 120 million ATPs; RNA, 6,000; and protein, 1,500 (see below, column n-ATP).

Table 7

molecule	Number of ATP molecules needed (n – ATP)	Number of molecules made each second (r)	Number of ATP molecules used each second (n – used)	Percentage of energy used to make the specified molecule (pb)
DNA	120,000,000	.00333	400,000	14%
RNA	6,000	12.5	75,000	3
Protein	1,500	1,458	2,187,000	78
Lipid	7	12,500	87,500	3
Sugars	2,000	31.25	62,500	2

$$(n-used) = r \times (n-ATP)$$

Given the replication time of the cell (20 minutes) and the number of constituents needed, we can estimate the rate, r, of molecular formation (how many molecules are made per second) by ATP energy. For example, to make 37,500 sugar polysaccharides in 20 minutes means that about 31 molecules are made each second. (The rates of molecular buildup are shown in Table 7 under the column labeled r.) With the buildup rate (r) and the number of ATPs needed (n-ATP), it is easy to calculate the number of ATPs used (n-used) in the synthesis process for each second that replication is going on. This is an important indicator of how quickly ATP is used up and, therefore, of how quickly it must be built up again.

Altogether, just under 3 million ATPs (2,812,000) are used each second to build a cell! Estimates show that only around 5 million ATPs are present. This means that the cell would die in just over 3 seconds if it wasn't possible to generate more ATP from ADP. When the cell is made to work hard, the rate of ATP production must also increase. Taking into account a reserve of around 1 million ATPs, this means that about 2 million ATPs must be made each second to sustain the processes of cell life.

From the above, it is also easy to estimate the percent of total biosynthetic energy, pb, used to synthesize each molecule. The formula is simply the ratio of n-used to the total number of ATPs used per second, or 2,187,000 divided by 2,812,000, which equals 78 percent (Table 7, column pb).

As you can see, the lion's share of energy goes to the construction of protein, with a reasonable amount directed to the buildup of DNA. These two processes take up about 92 percent of all the cell's energy for each second of cellular life. Thus, we see that body building requires most of the energy contained in the temporary storage vats of ATP molecules.

Somehow all the events I have described above fit together. Some organizing principle is at play, a principle that causes simpler units to form into more complex ones. To do so, biosynthesis goes against the grain of entropy, and to reverse entropy is no mean trick. Intelligence is needed, and so it is that intelligence manifests itself even at the level of the lowly cell.

CHAPTER 10

What Are Good Little People Made Of?

It is quite amazing to realize how much of our body's energy is used in rebuilding it. Most of the energy we use does its work at the quantum level—the level where the laws of quantum physics run the show. These laws work where atoms and molecules are interacting. To get a better understanding of body building, we next look at the atomic elements and how they can be formed into the carbon units that make up the living body.

The Elements of Life

The directions that spur the processes of life depend on information, information that is contained within the nucleus of the cell in the form of deoxyribonucleic acid (DNA). But how does a molecule of DNA contain information? How is that information actually used in the business of "is-ness"? Let us examine that question, keeping in mind the importance of quantum physics and the "consciousness" of quantum wave functions.

The Library of Congregated Atoms

We are composed of atoms. Atoms, in turn, are made up of heavy particles called nuclei consisting of protons, which are positively charged, and neutrons, which have no electrical charge but which have the same mass as protons.

Surrounding each atom are a number of electrons. Each electron, though nearly 2,000 times less massive than a single proton, carries an equal but opposite electrical charge, thus attracting it to the nucleus. Each electron also possesses an inherent spin and a magnetic field, called its "magnetic moment." The electron spin can be envisioned as a propeller spinning on its axis. A clockwise spin is said to be spin upward, a counterclockwise, spin down.

According to quantum physics, atomic electrons do not occupy a unique position at any given time. Instead, they appear in the form of quantum wave patterns which I call qwiffs. Although difficult to picture, you can imagine these qwiffs as spatially arranged in cloudy spherical layers surrounding the nucleus of the atom, much like the layers of an onion. These layers are called shells.

Electrons That Hate: Shells and Bonds

The outermost shell of each atom contains electrons that are able to form quantum physical bonds with other atoms. The strength of the bond depends on the number of electrons able to participate in its formation. The atoms carbon (C), nitrogen (N), oxygen (O), and hydrogen (H) are the primary elements found in each human cell, and they make up the set of 20 amino acids that are the building blocks of all protein.

The number of electrons available in the atom's outermost shells varies. The availability of these electrons enables atoms to form partnerships with each other called molecules. These links are energetically favorable; that is, when two or more atoms join together, energy is liberated. This joining "traps" the atoms into stable molecular ties. To dissociate the molecule and, thus, free the atoms would mean adding enough energy to break the ties. These ties, called quantum mechanical electronic bonds, are the "glue" that holds molecules together.

When an element has one electron removed, it becomes positively charged; when it acquires an additional electron, it becomes charged negatively. In each case, the element is said to be ionized, and it is called an *ion*. Ionization alters the kind of bond formed between atoms. For example, sodium (Na) and chlorine (Cl) are each in an ionized condition when they form the simple molecule NaCl, or common table salt. Sodium gives up one electron to chlorine. Thus, Na^+ and CL^- are each oppositely charged, causing them to be electrically attracted together as well as molecularly bonded. The bond between them is consequently quite strong, as evidenced by the very high melting point of salt—only at 1,473 degrees Fahrenheit is this bond loosened.

The ability to form electron bonds depends on the energy, the angular momentum (its movement around the nucleus), and the spin of each electron. These attributes comprise what is called a quantum state, consisting of four quantum numbers: the energy quantum number (N), the angular momentum quantum number (L), the projection of the angular momentum along a direction in space (M), and the spin quantum number (S). Taken together, they shape each atomic shell and ultimately determine how each atom combines to form molecular bonds.

To grasp how N, L, M, and S operate, it is useful to think of each electron as possessing a tiny unit of "quantum hate," keeping it apart from other electrons. This hate unit is composed, in turn, of the four quantum numbers: Each electron must possess a unique assignment of N, L, M, and S in every atom. For example, a hydrogen atom has one electron, with numbers 1, 0, 0, +1/2; a helium atom has two electrons, numbers 1, 0, 0, +1/2 and 1, 0, 0, -1/2. When we reach lithium we find three electrons; however, as only two electrons can exist in the first shell, the third electron must enter a higher shell. Here the "quantum hate" principle keeps that third electron from "playing" with the other two. Its complete quantum numbers must be 2, 0, 0, +1/2. Just why the universe should have a hate principle certainly is not clear. It seems to be connected with that last quantum number, ±1/2, called the electron spin. Whenever two identical particles each have a spin quantum number that is an odd digit divided by 2, such as 1/2, 3/2, or 5/2, they will never enter into a state in which all of their quantum numbers are identical. It's a kind of quantum identity crisis—hence, my use of the word "hate."

Thus, an electron will not enter any state already occupied by another electron. This is called the Pauli exclusion principle, and it explains why shells form in the first place. When a shell is filled or complete, the atom cannot accept another electron without forming a higher energy shell for it. When an atom has a complete shell, the atom resists the removal of an electron, and only lets one go with the addition of a relatively large expenditure of energy.

Sodium, in giving up one electron, finds itself with a complete outer shell. Chlorine, in accepting one electron, also completes its outer shell. This tendency toward shell-completion can be viewed as a driving force in molecular formation.

The inherent purpose of bond formation is the stabilization of a shell by giving it a complete number of paired electrons, each existing in a different quantum state. Two hydrogen atoms will bond with each other to form a molecule of hydrogen gas. The H atoms prefer marriage to the single state because their electrons overcome their hatred when they enter into an arrangement which groups them more or less in the same space. Since they cannot enter the same quantum state, they must alter their spins so that they are antialigned, head to tail. This compromise between electron hatred and energy favorability constitutes the quantum mechanical pact between electrons, enabling life through chemistry to exist.

80

Since carbon (C) has four available electrons, it can enter into a wide variety of bond formations, thus providing stable structures of three-dimensional molecular formations at quite specific directions in space, much like a Tinkertoy unit. (Indeed, chemists have used Tinkertoys as models for complex organic molecules made of carbon.) Each bond is composed of two electrons, one from each atom. Thus, a single carbon atom can form four single bonds because it has four electrons available. Or, it can form two single bonds and one double bond, or two double bonds.

It is this ability to form bonds linking it to four, three, or two neighboring atoms that gives carbon its utility as a building block of life.

Even though carbon atoms form stable bonds with other partners, they can also dissociate from them to form new partnerships with other molecules. How does this happen? In a previous chapter I described how enzymes worked. These pesky molecules are able to cause carbon-chained molecules to break their bonds by simply coming into their neighborhood. Once the bond breaks, the broken molecule temporarily bonds with the enzyme molecule and then flies from it: The enzyme bonding doesn't quite fit with its temporary carbon partner, so it breaks from it. The rates of reaction for these processes are extremely fast when enzymes are around but slow when they're not. With each breaking and remaking of a bond, electrons form new quantum pairs with electrons on new partner molecules. At even this level, intelligence—the quantum observer effect—comes into view. Carbon units appear to be reorganized by living conscious carbon units into living conscious carbon units.

All living molecules are made of carbon units. If it were possible to watch the processes of life taking place at the quantum level, where these bonds are continually regrouping and breaking apart, it would appear as a beehive of electronic magic. Individually, the bonds would be snapping on and off like a stroboscopic light show. It would appear maddening until one began to back off and watch from a distance. Only then would the intelligence of the quantum beehive begin to appear. Somehow, on this tiny, remote stage of space and time, consciousness is running the show and running it extremely well. From this vantage point, the body really shows itself as the body quantum.

CHAPTER 11

The Structure of Proteins

From the basic elements of carbon, nitrogen, oxygen, and hydrogen, 20 fundamental amino acids are built, which, in turn, make up the protein structure of the human body. These elements are also needed to build up the backbone of the master coder, DNA. It is within DNA that we find not only a mechanism for coding and arranging life's processes but also an interesting candidate for how miscoding, cancer, and death can occur. But before we get to this story, we need to look at the general structure of amino acids.

The Amino Acids—Building Blocks for Peptide Chains

The amino acid molecules are truly the molecules of life. They are amazingly versatile, and can be arranged into a great number of sequences. Since there are 20 amino acids, it is easy to appreciate how many molecular chains of protein can be built from them. Typically, proteins have 50 to 100 or more amino acids arranged like beads on a string. In a typical string of 75 amino acids, we find the outstanding number of over 3×10^{97} different possible arrangements.

LEFT-HANDED MOLECULES

One of the most interesting facts about amino acids is that they are left-handed! Amino acids can be thought of as "-handed" because each molecule, with the exception of glycine (GLY), can exist in two basic forms: the form found in human beings, and a mirror-image form. The mirror-image forms, called *sterioisomers*, do not exist in us but they can be constructed. No one knows why the left-handed form was chosen over the right-handed sterioisomer.

Adolf Hochstim, a plasma physicist, postulates in a journal article that the origin of handedness (referred to as *chirality*) is "probably associated with a difference in the initial concentrations of the two separate populations of primeval organic molecules and possibly even two types of primeval organisms and amplified by nonlinear kinetic processes leading to the death of one population."

In early history, when the earth condensed from stellar matter, the planet was quite hot and radioactive. After a long period of cooling, organic molecules were produced in the primeval atmosphere. A number of processes account for this production, including lightning, ultraviolet radiation, quenching reactions from shock waves due to falling meteorites, lightning on water surfaces, and shock waves from thunder. There appears to be no way to explain from these processes how L (left) molecules would appear in any greater number than D (right) molecules. (The letters L and D stand for *levo* and *dextro*—Latin for left and right.)

Thus, both L and D amino acids must have appeared in the primeval "soup" in approximately equal number, subject to the usual fluctuation. Even if there were asymmetry in this early phase, slow conversion of L to D or D to L would have transpired in the relatively short period of 100,000 years.

Therefore, organisms must have appeared with both L and D amino acids. Surely these organisms must have been able to live, reproduce, and die. Suppose we label an L-organism as L and a D-organism as D. Processes exist which enable L and D to clone, so that one L produces two Ls at a given rate and one D produces two Ds at the same rate. At the same time, L and D can each die at identical rates, different from their rates of reproduction. Obviously, if the death rate is greater than the reproductive rate, the organisms will eventually die out, but if the reproductive rate exceeds the death rate, they will survive and multiply exponentially. If the initial concentrations of L and D differ due to a fluctuation, then that difference will also grow in time, with one exceeding the other.

If, on the other had, there was a process where a simple pairing of an L and a D resulted in death to both organisms with a killing rate (caused by one organism eating the other and suffering the worst kind of indigestion, death, or having the two fight and kill each other off), then the process of survival becomes more complex. We now have three processes:

> L reproduces or dies;
> D reproduces or dies; and
> L and D kill each other.

The processes are interesting because of the complexity introduced by the antagonistic behavior. This results in one species always surviving and, indeed, growing exponentially at the expense of the other, which becomes extinct, even though both die on encounter. This always occurs if one species exceeds the other initially. However, if both species exist in equal numbers to start with, L and D approach an equilibrium, a balance—a kind of predator-prey ecology.

It appears likely that this complex or nonlinear antagonism could have produced the left-handed amino acid organisms surviving on our planet today. Perhaps research on other planetary life forms will show that D-amino acid life forms exist. It would follow, too, that on any given planet only a D- or an L-evolved life form would be present, not both. What would happen if earthlings were to encounter evolved D-life on some distant planet? Would there be a natural antagonism between the two species? If so, this would be the first case of quantum mechanical molecular war.

Why Amino Acids Are Acids and Why They Are Amine

The three-dimensional distinguishability of an amino acid from its mirror image is important in the way that it interacts with other molecules. If you were to build up a Tinkertoy replica of an amino acid, there would be no way to transform it into its mirror image without taking the toy apart. The key to understanding amino acids (even including why they are called amino acids) lies in their end-pieces. On one end is a molecule called amine containing two hydrogen atoms stuck out in space from a nitrogen atom so that the amine forms an equilateral triangle. However, this simple structure doesn't just sit there in space: The H atoms are quantum physical in their attachment to nitrogen (N), and they tend to exist potentially pointing both toward the left and right of N with equal probability. We will return to this quantum mechanical fact later on. Since all amino acids have an amine left side, so the designation *amino*.

The right end of the amino acid is where the name acid comes from. It consists of a carbon atom double-bonded to oxygen, above and to the right, and single-bonded to a hydroxyl molecule (OH), below and to the right. The hydroxyl consists, in turn, of a single hydrogen atom in a single bond with oxygen. This unit, COOH, called carboxyl, is the most common feature of all organic acids.

Carboxyl is generally an acid because it can dissociate (come apart) in water, forming both a proton and a negatively charged carbon dioxide ion. Acids are compounds that produce protons in solution.

The amine, on the other hand, is considered a base (the opposite of an acid) because it tends to pick up protons in solution.

Thus, the amine end of an amino acid tends to attract positively charged hydrogen atoms (protons), while the right-hand side tends to give them off.

Amino acids come in three general classifications: water-fearing, water-attracting, and charged. Another way to put this is nonpolar, polar, and charged, respectively. The nine water-fearing or hydrophobic amino acids are the molecules GLY, ALA, VAL, LEU, ILE, PRO, PHE, TRY, and MET. This means they do not bind to water molecules. If you have ever tried to dissolve carbon in water, you will see what hydrophobia is all about. The old adage that oil and water don't mix is due to the water phobia of carbon. There simply is no way for water, which is polar (has an electric field), to hook up with carbon conveniently. The fact that some amino acids are water-fearing is very important in the functioning of the protein layer separating cells.

A polar, or water-attracting, molecule is one that, though electrically neutral, generates an electric field surrounding it. A charged molecule, of course, contains one or more electrical charges. The main key with polar amino acids is the presence of the hydroxyl (OH) or a single oxygen that attracts polar molecules. The six polar amino acids—SER, THR, CYS, TYR, ASN, and GLN—all are electrically polarized, and thus are able to bond with water molecules.

The five remaining chargeable amino acids are neutral molecules that assume a charged state when immersed in a solution which is considered to be pH neutral. (The pH refers to the concentration of free protons found in a solution. Acids have lots of protons, so they are said to be pH low; water is pH neutral; bases are pH high, with an absence of protons in solution.) The charged amino acids—ASP, GLU, LYS, ARG, and HIS—are all able to bond electrically with other molecular structures.

The Structure of Proteins

All proteins are built up from amino acids. In the human cell, the ribosomes carry out the function of chaining together the various amino acids into the proteins needed by the body. The two ends of the amino acid groups are the same in every case. (Note that although proline doesn't quite fit the scheme, it still contains one nitrogen atom and two hydrogen atoms at the left end.) These ends link when one amino acid is joined to another. With each link, a two-electron bond forms between neighboring carbon and nitrogen atoms.

Through the actions of quantum physical forces, complex protein molecules were built up from carbon atoms arranged in these groups of amino acids. Here we see quantum physics acting as the master builder. Our early history shows that these amino acid molecules were created left-handed, capable of attracting water, repelling water, or behaving neutrally toward its presence. From just 20 amino acids all of the proteins used to build the body are constructed.

The main problem with all this is that over 200,000 proteins have been constructed for use in our cells. Since each of these proteins is, in itself, a complex arrangement of the 20 amino acids, it doesn't take long to figure out that the number of possible arrangements is enormous. Fred Hoyle, in his book *The Intelligent Universe*, estimates that to make just one of these proteins requires enormous odds.

Think of each amino acid as a bead, and each protein as a string of these beads. Remember that a typical protein is akin to a string of about 100 beads, containing, at the most, 20 different colors. How many possible arrangements are there? The answer is 20 raised exponentially to the power 100! That's 20 multiplied by itself 100 times. The enormity of this number makes for such small odds that any particular protein will be built that we must rule out the possibility that proteins were constructed from random or nonintelligent forces. The only answer facing us is that intelligence was at work then, and that it is still at work today. The body quantum, through the observer effect, was choosing back then in the dark halls of our early history. Who was the *observer*?

CHAPTER 12

The Quantum Mechanical DNA Coding Machine

Building proteins requires intelligence; that intelligence is contained in the complex molecules known as DNA. Many of the explosive advances in molecular biology came from the realization that DNA had a spiral-staircase structure, which became apparent through the advent of X-ray crystallographic studies. The latest work in molecular biology is aimed at understanding how DNA interacts with other components of the living cell and how DNA expresses its information to those components. In this chapter, we will look at some of the details of the structure of DNA; later, in Chapter 29, we will need to recall this information.

My reason for including it here is to show briefly how the DNA molecule acts as a coded library book, the code containing the instructions for building a human being.

How DNA Is Built

DNA is a nucleic acid, so named because it is found in the nucleus of the living cell and consists of a backbone made of a sugar molecule and a phosphoric acid. Through the work of Francis Crick, James Watson, Maurice Wilkens, and many others, the complete structure of DNA was determined. It consists of two snake-like strands winding about each other in a continual spiral, as if up a pole. Connecting the strands are small molecular groups called bases. These bases are changeable and, therefore, capable of containing or carrying information. The bases tend to form steps between the strands, each step containing a pair of bases in a complementary arrangement.

There are two kinds of sugars in nucleic acid, ribose and deoxyribose; and each sugar molecule is in the shape of a pentagon.

The remaining jigsaw piece of the backbone building is phosphoric acid, simply written as P. It consists of a single phosphorus atom surrounded by four oxygen atoms, with three of the oxygens attached to hydrogens. (Remember, an acid is defined by its ability to produce protons or hydrogen atoms with electrons removed.) Phosphoric acid, or P, has three protons to give up to a solution and, therefore, is quite acidic. (An acid stomach is full of free protons; you might say your acid stomach is a proton bag.)

The Backbone of Life

To form a backbone for DNA, we alternate sugar-P-sugar-P, etc., in a chain. The pentagonal rings of deoxyribose are attached together by the phosphoric acids forming into a phosphate. One might say that the backbone of life is an unflavored sugar phosphate—an egg cream, to East Coasters! It's just one sugar phosphate after another, end to end. Some atoms are missing: The phosphoric acid has lost all four of its Hs and one of its Os; even the sugar has lost a hydroxyl, OH. Adding up all the losses gives a pair of water molecules and a proton, making the DNA backbone negatively charged.

So Where's the Message?

The final pieces making up the DNA molecule are the five distinct bases: two of these bases are called purines, the remaining three, pyrimidines. Each base tends to lie in a flat plane. Each purine consists of a single hexagonal ring cojoined with a pentagonal ring, while each pyrimidine consists of a single hexagonal ring. Each base is capable of attaching itself to the backbone of a nucleic acid. To DNA, the bases adenine (A), guanine (G), cytosine (C), and thymine (T) will attach, while to RNA, its sister molecule uracil (U)—which acts as a transcriber and a messenger of the DNA code instead of the chemically similar thymine (T)—is attachable. Thus, RNA can take on the four bases A, G, C, and U, while DNA will take on bases A, G, C, and T.

Herein lies the means for writing and transcribing messages. These bases are the alphabet of the DNA code. Each base can only connect to a complementary base. Thus, C will connect with G, but not with A or T; A will connect with T, but not C or G. In this way, a molecule of DNA is made from two winding chains in a spiral staircase, with complementary bases forming the steps. The rules which govern the bonding of bases are ultimately the rules of quantum physics. Through these rules, life is able to carry out the intelligent process of building the body.

Each base attaches itself to a particular position on the sugar ring. What makes life interesting is the arrangements of the bases and the fact that the bases will only connect with each other in certain ways. A unit consisting of a sugar phosphate and base is called a *nucleotide.*

Between bases there exists a set of subtle quantum mechanical bonds that, I believe, are the keys to longevity, health, and quantum mechanical diseases such as cancer. These are hydrogen bonds and correspond to a single proton trapped between negatively charged atoms.

Two hydrogen bonds will form either between A and T or between A and U. However, A will not find hydrogen bonding favorable with C because there appears to be no favorable angle for all three bonds to form. It is G that bonds with C, forming three hydrogen atom bonds. G has two hydrogen atoms to give in forming two of the bonds, and one to take from C in forming the third bond. G will not bond with U or T, because U and T each have only one hydrogen atom to give and one to take in forming bonds. Thus, either U or T ideally matches with A, which also has one hydrogen atom to give and one to take in forming two hydrogen atom bonds.

The reason for this selectivity in partners is again quantum physical. A, thus, can form two bonds, while G can form three; U and T can form two bonds, while G and C must form three. This leads to the pairings between the bases: $A = T$ or $A = U$ and $G \equiv C$. The double lines between A and T and between A and U signify the double electron bonds between them. The triple lines between G and C indicate that a triple electron bond forms between them.

Since a typical DNA molecule can have as many as 200,000 base pairs linked together, it is now believed that the coded message for life lies within these possible arrangements. The specificity of A with T and G with C also tells us how DNA replicates itself.

You and I are made of proteins built up by chains of amino acids. The rules for building a protein follow the templates of the veritable library of DNA. The books in the library are written with four-letter alphabets. From these are built humans who speak with a variety of different alphabets, forming the human languages. The alphabet of DNA's bases ultimately produce the alphabets of humanity.

To grasp how this is possible, think of the DNA molecules as library books written with long, complex words made from a simple four-letter alphabet. The writer of these books lived a long time ago, and yet is very much alive today! Think of the RNA molecules as little students who are able to read these books and form copies of the words. The only difference is that the RNA "students" use as their alphabet the letters A, C, G, and U, while books of DNA are written with the letters A, C, G, and T. However, these RNA students are actually created when they enter the nuclear library and read the DNA books. They are little golems. At this level, inside the cellular nucleus, we have a group of very impressionable students, to say the least. Like good students everywhere, they will travel to other parts of the cell, as I shall point out in the next chapter, where they act as life's intelligent managers and messengers.

CHAPTER 13

An Overview of a Living Cell: Information and Protein in Motion

To gain a better understanding of how our bodies are built, we need to look at how a living cell manages itself. A key ingredient is not the food itself, but information, and how that information is processed.

Inside a Cell

The cell, the smallest unit of life, carries out a number of complex operations. We have a good understanding of these operations, and we even know how a cell follows its own instructions and where those instructions are generated and stored. Yet just how a cell "knows" how to perform those operations remains a mystery.

In certain top-secret organizations, information is channeled according to individuals' "need to know." Does a cell have a need to know its own processes? When does a cell become part of a conscious, knowing unit? At the level of organs? Brains? Nervous systems? Or perhaps cells possess their own form of cellular consciousness. So far, no one knows.

We do know that the cell needs no outside intelligence to tell it when and how to do its work. The secret of cellular intelligence lies within the tiny nucleus inside the cell itself. In fact, the nucleus is so important that cell biologists have divided all cells into two types according to the presence or absence of a nucleus: *prokaryotic*, or *eukaryotic*; terms derived from the Greek *karyon*, which means kernel or nucleus.

Prokaryotic cells are those which may be in the stage of forming a nucleus, while eukaryotic cells are fully developed with nuclei. (In this case, the prefix *pro-* means before, as in prototype.) Bacteria and blue-green algae are examples of prokaryotic cells, able to function quite nicely despite their lack of nuclei. They live by interacting with each other, exchanging materials and even working together as part of a larger system toward some common goal or aim.

It is even possible that the mitochondria in the human cell are prokaryotic cells. In other words, the mitochrondia, the organisms within each human cell that are responsible for the oxygenation of foodstuffs and the production of ATP, may be thought of as cells living within cells.

Prokaryotic cells are accommodating creatures, able to propagate, reproduce, and eat without a nuclear library to tell them what to do or how to do it. Possibly, these simple cells carry out these processes using the environment as their intelligence. They function according to the world they live in.

By contrast, eukaryotic cells, or eukaryotes, contain DNA within their nuclei and thus carry their own library of information with them. Undoubtedly, this enables eukaryotes to carry out more complex tasks than prokaryotes. Comparing them is somewhat like comparing a book-laden scholar to a primitive cave dweller.

Human cells, without exception, are eukaryotic cells. Each cell is able to maintain, repair, and multiply itself—the three functions considered to be the primary functions of living stuff. Each cell is both a quantum and a classical mechanical system, and so both classical and quantum physics are needed to understand how a human cell functions.

The wall surrounding a cell is some 60 millionths of an inch thick. It is made up of a double layer of fat molecules called lipids. Embedded in this wall of fat are protein molecules, separated from each other by only a few billionths of an inch. These proteins act as gateways for the cell to exchange oxygen and carbon dioxide; nutrients and waste products also pass through these gateways.

The cell can be pictured as a factory. The cell nucleus works like a computer organizing the cell's productive capacity. Like a factory, the cell, in a continuous flow, takes in raw materials for processing, and disposes of its own waste products through the cell wall.

Directing those processes is the function of the nucleus; being directed are the ribosomes, lysosomes, mitochondria, centrioles, Golgi complexes, endoplasmic reticula, cytoplasms, and nucleoli.

Continuing the factory image, think of the cell as containing different manufacturing units. Each unit must carry out a specific function and produce a specific product. These units must also communicate with each other.

Directions for production come from the cell nucleus. The nucleus also receives information from the manufacturing units surrounding it. It acts as a communication center. When incoming information is received, messages are then generated and transmitted to the rest of the cell. These messages are units of intelligence contained in the nucleic acids library.

The different manufacturing units within the cell produce substances needed for cell maintenance and energy. These substances are called *loads,* and consists of amino acid building blocks. The assembly of these blocks into actual amino acids constitutes the work of the factory.

As I mentioned earlier, there are two types of nucleic acids used in cells. These are deoxyribonucleic acid, or DNA, and ribonucleic acid, or RNA. RNA is of two types: messenger RNA (mRNA), and transfer RNA (tRNA). The mRNA carries messages, while tRNA carries loads to be transferred.

The Endoplasmic Reticulum: Channels for Communication

The endoplasmic reticulum is a network of flattened sacs permeating the body of the cell and providing the channels for communication from the nucleus to the rest of the cell. The sacs contain on their surfaces structures called *ribosomes.*

RIBOSOMES: FOLLOWING INSTRUCTIONS FROM MRNA

Ribosomes are the true protein manufacturing units of the cell. To briefly explain, they receive their manufacturing instructions from mRNA. The mRNA looks like a long ticker tape threading through the ribosome. Embedded along the length of the mRNA is a coded message. The ribosome reads a section of the message, then moves on along the ticker tape to read another part of the message.

Reading consists of matching the string of molecules stretched along the length of the ticker tape with a complementary string of tRNA. The tRNA can be compared to a truck carrying a load, a truck that is color-coded. The ribosome opens its doors to a molecule of tRNA. If the color code complements the piece of the message carried by the mRNA ticker tape threading through the ribosome, the tRNA remains in the ribosome; otherwise, it leaves.

To picture this, imagine a ribosome as a subway station in New York, with two trains on parallel tracks stopped in the station for loading and unloading passengers. One train is the tRNA "express," the other, the mRNA "local." The problem is that the tRNA train conductor, who controls the train's doors, will not open them and discharge amino acid passengers at any time. He looks at the other mRNA train. If the car from the mRNA train just opposite him is the right color, that is, if the word written on its side (by some intelligence, we hope—we can only guess that it is neither advertising nor graffiti) is the right word, the tRNA unloads its amino acid "passenger." If not, the tRNA train moves out and another comes in. This time the tRNA conductor is different; his train carries a different amino acid "loaded" passenger. He will be looking for a different word. If he sees the right word, he opens the door and an appropriate amino acid passenger departs, climbs the stairs, and joins a train just waiting for him and carrying other amino acid passengers. The tRNA train leaves and another comes in as the mRNA train moves up one car along the track, where a new word becomes visible to the opposite train.

Actually, the process is just slightly more complicated.

Once the codes match, the tRNA carries out a transfer operation involving itself and a second tRNA molecule; both must match codes with an adjacent piece of the mRNA. The second tRNA departs the ribosome and discharges its already existing chain of amino acids. The first tRNA then switches over carrying its load of a single amino acid, and attaches it to the rear end of the already existing chain of amino acids. When the ticker tape moves through the ribosome, it transfers the tRNA along with it, making room for another tRNA to enter the ribosome. The previous operation is then repeated. This continues until a chain of protein is built up in the ribosome.

LYSOSOMES: CHEMICAL STORAGE LOCKERS

The lysosomes are simple sacs storing enzymes which are used in the digestion of nutrients: The cell actually eats. These enzymes break down large or long protein chains into smaller units, which are then utilized by the cell to make other protein chains. They are a kind of recycling plant.

MITOCHONDRIA: ENERGY TRANSFORMATIONS

The mitochondria are the main sites for energy production. In these sausage-shaped structures, which are about 1/1,000th of an inch long, are found elaborate folds of membrane. Here food is oxygenated and the energy units of adenosine triphosphate (ATP) are produced.

The energy of ATP is then utilized when the ATP delivers its electrical energy load to the various parts of the cell needing it. This energy is contained in the quantum mechanical electronic bonds as potential energy, residing between phosphorus and oxygen in the molecule of ATP.

ATP is one of a family of compound molecules known as *nucleotides*. In the living cell, ATP is also highly ionized, containing four charges of electrons (its pH is 7.0).

CENTRIOLES: MAKING SPINDLES FOR SPLITS

The centrioles are involved in the process of cell division or mitosis. They move apart when the cell is about to divide. In so doing, they leave a delicate trace of themselves, called spindles, throughout the cytoplasm (the cell body). The exact way this all happens is not well understood even today.

GOLGI COMPLEXES: DEPOTS FOR NEW AND OLD PRODUCTS

The Golgi complexes serve as depots for the storage of cell products and for their secretion. They are gatherings of thin, flat, saucerlike bodies; they take in protein polypeptide chains, produced by the ribosomes, and attach carbohydrate units, forming what are called glycoproteins.

CYTOPLASMS: THE MEDIUM OF THE CELL

The cytoplasm makes up the main body of the cell; it is the watery medium in which all the other structures float. The nucleus is separated from the cytoplasm by a very fine, semiporous membrane, allowing certain molecules to pass back and forth. Also floating within the cytoplasm are spherical globules of fat. (Within the body, there are special fat cells (so-called), found in the adipose tissue surrounding liver cells, for example, or just under your skin.)

NUCLEOLI: THE BUILDING BLOCKS FOR NUCLEIC ACID

The nucleolus is found within the nucleus and contains concentrations of RNA needed for communication to the ribosomes.

Cellular management depends on both the correct utilization of raw products, consisting of amino acids and other molecules, and a complex communication network. All these messages are contained in molecules. Since molecules are themselves composed of atoms and since they are held together by quantum mechanical forces, perhaps the key to the intelligence of the cell can be found within the molecules themselves.

It is difficult, perhaps, to realize or even to conceptualize molecules containing or exhibiting intelligence. Does the intelligence

95

we seek in such microscopic entities appear as a mere projection of our own intelligence? I don't think so, because all the remarkably intelligent processes that take place in our cells are relics of the past. Already we have seen that even our distant molecular ancestors—our primordial proteins—could not have been assembled just randomly.

Some form of DNA-RNA replicating "library with students" must have appeared when life itself appeared. If so, then where does that intelligence reside? It may be that this intelligence comes from some Father or Mother in the sky. But I believe that the universe works in a more subtle manner.

The God of old must have known quantum physics to construct the living paradox we call human beings, and all other living entities as well. It is through choice that cellular management and evolution can work their wonders, and it is through quantum physics that we see how that choice is manifested. It is again the body quantum—the observer effect in action—that, by choosing the improbable every once in a while, makes life different from mechanics. Through such actions, either success or failure is the result, and that which succeeds results in new life, perhaps even magic on the landscape of existence.

CHAPTER 14

Reproduction: The Quantum Physics of Sex

Human beings are probably the sexiest creatures around. Flowers do it, animals do it, but Homo sapiens has taken sex out of the biology textbooks and put it into the mind. What makes us so sexy? Probably it is because we are more conscious of our bodies than other animals. If we could become even more conscious, we might become aware of the cellular divisions and cellular deaths taking place within.

Are we not becoming more conscious? It seems to me that building up and death are closely related. With the evolution of consciousness, these processes—cellular splitting, sexual cell production, and cellular death—would be felt. Perhaps the awareness of these processes would constitute a form of cosmic consciousness: that gray area that separates life and death.

The drive to couple sexually probably stems from a deep-rooted and fundamental "feeling." This feeling really isn't an ordinary one that we are accustomed to; it is far more basic—more like a certain tendency—that exists at the molecular level of our being, causing changes in our bodies at the gross level of expression. The molecular feelings arise from the body quantum. It's the observer effect once again choosing among probabilities. This time the effect can be thought of as a reaching out—a desire for something beyond immediate awareness. That reaching out feeling manifests at our conscious level as sexual desire; fundamentally, a desire to create beyond the limits of our own skins. In so doing, the laws of quantum physics can be magnified: New structures will appear—our children. From sexual bonding, the universe manifests life forever; another form of building life takes place.

Cell Division

Every minute in every human body some 300 million cells die. If cells were not able to reproduce themselves through mitosis, every cell in the body would be dead in about 139 days. (The body contains 60 trillion cells. Divide that by 300 million cell deaths per minute, and you get 200,000 minutes. Divide that by 60 minutes per hour and then again by 24 hours per day, and you arrive at 139 days.)

During the first two steps of mitosis, the nuclear membrane breaks down, causing the release of chromosomes throughout the whole cell. Chromosomes are coiled rods of DNA molecules residing within the cell nucleus. Each human cell nucleus contains exactly 46 chromosomes, with the exception of sperm and ova cells, which contain 23 chromosomes each. Chromosomes were discovered shortly after Gregor Mendel's discovery of the field of genetics; they are the cellular bearers of heredity.

In 1866, Mendel discovered the essential features of cellular inheritance: An offspring acquires its characteristics from its parents through units that now are called genes. Soon afterward, sperm and ova cells were observed to contain little cytoplasm, or cellular fluid, in relation to their nuclei. Cellular nuclei then were inspected, using special dyes, and were found to contain units that absorbed the dye. These units were called chromosomes; from the Greek root *chromo*, meaning color. Cells of members of a given species were found to include a constant number of chromosomes. When examined closely, by the use of electronic micrographs, each chromosome was observed to have a characteristic shape.

The first trait assigned to a chromosome was sex itself. Every cell of a given species, although appearing nearly identical, exhibits either male or female characteristics. Thus, individual cells from males are different from female cells. That difference appears in only 1 tiny chromosome contained within the cell nucleus. A careful examination shows that there are 2 sex chromosomes; they are shaped in the forms of the letters X and Y, with Y being the shorter of the two. If the cell nucleus contains 2 X chromosomes, it is a female cell; if it contains an X and a Y, it is a male cell. No cell contains 2 Y chromosomes.

A single chromosome contains around 500 different DNA molecules, or 500 different genes. Thus, each huge molecule of DNA is a gene—the units that taken all together comprise instructions for all of the bodily characteristics we observe. Taking 500 genes per chromosome times 46 chromosomes yields 23,000 different possible characteristics for such things as the body's build, its inherited instincts, the color of skin and eyes, reaction times, height, and even possibly the way the body will eventually die.

Since we live beyond 139 days, our cells must divide, and after division, each must contain the same library of genetic material the original contained. In other words, each cell must contain the same number of chromosomes as the original cell, thus making each offspring cell a duplicate of the original.

When a cell is about to divide, each DNA molecule splits apart: The double-helix structure "unzips." This step actually encompasses the first of four distinct phases of cellular division: the prophase, the metaphase, the anaphase, and the telophase.

During the early prophase, the chromosomes appear denser and more condensed. The cell's nucleolus starts to fade, and the centriole separates into two, forming centers of radiating filaments attached to a spindle. During the later prophase, the chromosomes are even more condensed, with each doubled and connected (or joined) at the centers. There are now 46 pairs of chromosomes, or 92 chromosomes in all.

During the metaphase, the nuclear membrane dissolves and the chromosomes "invade" the cytoplasm. The centrioles separate further to opposite poles (north and south) of the cell. Radiating from each pole, they form filaments; in between the poles, the chromosomes are lined up to form what is called the *metaphase plate*. This plate cuts through the equator of the cell. The northern and southern sides of the plate contain duplicate members of each pair of chromosomes. Thus, one member is ready to migrate northward, while its twin migrates southward. The larger chromosomes are found on the perimeter of the plate, while the smaller ones are in the center.

With anaphase, the pairs of chromosomes begin to separate and start to move toward the opposite poles. The cell begins to elongate.

When telophase commences, the cell lengthens more, with the chromosomes well on their way toward the poles. With late telophase, a furrow appears, moving along the diameter of the cell. The nuclear membrane re-forms around the chromosomes, and the cell splits apart.

Consider a hypothetical cell containing 4 chromosomes: F, H, M, and W. The letters F and H refer to genes that have come from the father, and the letters M and W refer to genes contributed by the mother. Chromosomes tend to form in pairs—one member of the pair coming from each parent. Thus, 2 pairs might be F-M and H-W. When a cell undergoes mitosis, each chromosome is duplicated, producing FF-MM and HH-WW. The copies then are split apart and separated, with one copy going to one centriole pole, the second toward the other pole. The parental pairs then reassemble once again. When the cell divides, each offspring cell contains an exact replica of the chromosomes F-M and H-W found in the original cell.

Sexual Cells

The end product of mitosis is 2 exact replicas of the splitting cell, each containing 46 chromosomes. However, there is another process, called *meiosis*, in which the end product is not 2 identical twin cells with 46 chromosomes each, but 4 identical *gametes*, each with 23. These gametes are the sex cells that can combine to form a single cell with 46 chromosomes, 23 from each parent gamete.

Each functioning cell can be seen as composed of 23 pairs of chromosomes, making up the 46. Although each partner of a pair is not identical, the general size and shape of each partner is similar, only the sequences of genes in each differ. This is true for every pair of chromosomes in every cell except the male sex chromosomes. Male nuclei contain, instead of a matched pair of sex chromosomes, a hybrid pair consisting of an X and a Y. Female nuclei, on the other hand, consist of 23 pairs of matched "shape" twins. Thus, when a male gamete or sperm cell is produced, containing 23 chromosomes, it will contain either an X chromosome or a Y. When a female gamete or ovum is produced, it will always contain an X chromosome. Since males have XY pairs in each cell, their gametes will contain, with a 50 percent probability, either an X or a Y. Thus, fathers produce both male and female gametes, while mothers produce female gametes only. Mothers make only "daughters," while fathers breed "sons" and "daughters" in equal number. In this manner, when a sperm penetrates an ovum, the zygote, or resulting cell, has a 50 percent probability of being either male or female.

Now consider a hypothetical cell undergoing the process of meiosis. The first stage of duplication is the same as for mitosis: F-M and H-W become FF-MM and HH-WW. However, another possibility emerges for a meiotic cell—the possibility of rearrangement. Thus, FF-MM and HH-WW become FF-MM and WW-HH. In the rearranged possibility the WW changes place with the HH. The father-chromosome F matches with the mother-chromosome W. Another randomizing occurrence also can take place—the phenomenon of crossing. Here, parts of duplicated chromosomes "cross legs," so that each chromosome pair contains a hybrid of both parents: FF-MM becomes FM-FM, and WW-HH becomes WH-WH. With rearrangement and crossover, a different combination of hybrid chromosome pairs is generated.

Splitting, separation, and first-cell division now result in four different cell possibilities, of which either the first two or the second two actually are produced. With the next split, producing gametes, we find eight possible gene combinations, of which either the first four or the second four actually are produced. Thus, with a large number of sex cells, all eight gamete possibilities will appear.

In an actual human meiotic cell, there are 23 pairs of chromosomes, which means that 2^{23} rearrangements are possible—over 8 million different gametes! Including all the possible crossovers increases this number nearly beyond count. The purpose of meiosis thus appears to be a means for randomizing the process of creation.

The Quantum Game of Sex

As we have seen, the function of mitosis is the production of clone cells—each cell an identical copy of the parent cell. Meiosis, on the other hand, is the antithesis of mitosis. The game of the meiotic cell is to produce gametes with as many possible differences as can be created. The mechanism for this production is crossover and rearrangement. If meiosis occurred without these two processes, we would have, instead of eight possible, all-different gametes, four gametes in two identical pairs or clones. Suppose that these gametes were from a male. That would mean half of his sperm would contain cells with his father's characteristics and the remaining half, his mother's. If his sperm were to combine with a similarly generated genetically endowed clutch of ova, we would find all of the offspring appearing as identical replicas of just three types of cells.

Through what process does sex arise? I believe that sex arises as a direct consequence of quantum physics. Its randomization processes occurring at the level of sex cells are caused by the natural principle of indeterminacy inherent in all of the universe.

Biologist E. H. Mercer, in his book *The Foundations of Biological Theory*, takes a similar view. He offers an explanation of how the randomness contained in quantum physics is amplified or developed, and thus manifested at the macro level of cellular and human existence.

Consider a simple wheel, rotating with an angular velocity ω. This means that it turns through one complete revolution every

6.28/ω seconds. The number 6.28 is the angle of complete rotation in units called radians; it is the same as 360 degrees. The angular velocity, ω, is the number of radians turned per second.

Now, according to quantum physics, the velocity of the rotating wheel can be known only within a particular uncertainty, μ. This uncertainty may indeed be quite small. Nevertheless, the true angular velocity will be (ω + μ), the perceived angular velocity plus the uncertainty.

It may seem confusing to add an uncertainty, μ, to a certainty, ω. The point is that through the uncertainty principle we really don't know that the angular velocity is precisely ω. It could be as low as ω − μ, or as large as ω + μ, or any speed in between. That's the whole point of the uncertainty principle.

Suppose we wish to determine the position of the wheel—the angle in which it is to be found—after many, many turns. Since the wheel turns through a circle, it repeats its position every 6.28 radians (or 360 degrees). If the wheel turns for t seconds, the position will be the product of the angular velocity times the time, (ω + μ) × t. The term ωt is the expected result. But the error term μt can be quite large because t can be quite large. Since the wheel is rotating and repeating its position, the error becomes a sizable angle.

Perhaps, in a similar manner, errors at the quantum level are amplified at the meiotic cell level, and, thus, are responsible for the differences in gametes that are so necessary for the species to evolve into novel life forms. These quantum errors can give rise to both evolved characteristics and, regrettably, to throwbacks, "bad seeds."

To understand how this might occur, we need to look at quantum physics from a slightly different perspective. From this view, atoms, molecules, and subatomic matter really exist only as probability patterns—patterns of our minds, or of the One Mind that composes us all. From this view there is no objective reality in the present moment, and certainly none in either the past or the future.

This opens up the whole question of evolution. Toward what are we evolving? If we look at evolution as a logical progression to a specific goal in the future or as a consequence of actions in the past, I am afraid that both views are mistaken.

Imagine that, starting from the present, there exists a frontier of future possibilities. Possibly, in some of them people have hair growing on top of their noses, in others, they do not. Also imagine that there exists, simultaneously, a whole rear guard of possible pasts, all leading to this present. (Through scientific investigation, a consistent picture of the past can be made.) In some of those pasts, humans had hirsute noses; in others, they did not. Humans did not have hairy noses way back then. However, as we learn more, that picture of the past keeps changing, sometimes only slightly, while at other times radically. Only *some* humans had hairy noses, we then discover.

In a similar manner, the future consisting of several possible alternative futures spreads out before us. The reason for this multiple future is meaning. From a quantum point of view, it takes two events to have anything meaningful take place. A single event needs to be repeated in the future in order that the event can be related. That is why science insists on repetitive experiments. Similarly, past events are considered meaningful if we can somehow reproduce them in the present.

Somewhere in the future there exists a tribe of humans with hairy noses. There also exist tribes without them. The so-called throwbacks or genetic errors that crop up from time to time are nature's way of checking for a self-consistent agreement between a possible future and a present moment. If people begin to wish for hairy-nosed children, they will tend to "tune" to the hairy-nosed future, and then influence the possible resonance with that future. Most of us want children with hairless noses. So evolution tends to move in that direction. This effect can be attributed not to any mechanical influence in the present but to our collective dream as a human race.

Genetic diseases, such as AIDS, will crop up from time to time because there exists a possible future of human life in which that program produced by AIDS has taken a strong root. The problem is that through the gambling wheel of quantum life anything can occur, any future is possible. We must be careful what we wish for, even unconsciously.

An Insight

One of the most important ideas in the cosmology of the new physics addresses our deepest human concern: Why we are here, and what are we here to do. After much study and contemplation, I have come to believe that human life has only one purpose: to awaken itself and to awaken all matter. That is why evolution has moved toward increased intelligence and consciousness. Mitosis could never have done the job alone. It produces only clones. It appears that meiosis, and hence sex itself, was necessary to produce variety and to give new evolutionary combinations a chance.

However, as Charles Darwin wrote in 1862, "The whole subject is as yet hidden in darkness." That darkness still exists to some extent. Sex is defined by most biologists as "the process whereby a cell containing a new combination of genes is produced from two genetically different parent cells." The problem is, as molecular geneticist Norton Zinder put it, "How could an organism that only passed half of its genes to its offspring [through sexual reproduction] ever have competed with [an asexual] progenitor that passed all of them?"

Zinder feels that an asexual progenitor passing on all of its acquired and inherited characteristics would have given its children more of a chance to compete. It wouldn't have to take the chance of mating with a partner that had possibly poorer characteristics.

The answer, evidently, is that nature likes the long odds. In other words, it is willing to take the chance of a possibly more fit progeny by mating than the possibility of passing on a set of fixed characteristics from a single parent.

My insight goes beyond the classical biologist's. I believe that through quantum physical interactions, molecules in the parent sex cells "sense" each other's presence. This sensing most likely occurs during the fertilization process, just before the sperm enters the egg. It leads to correlations between the parent sex cell molecules. If that correlation tends to produce a fitter offspring, the probability will be enhanced that the offspring will be produced. If the correlation produces a less fit offspring, the probability will be decreased.

PART FOUR

Sensing

As World War II broke out, scientists began to work on a serious problem—the detection of invading forces by land, sea, or air. In ancient times, sentries, posted in strategic places, sent warning signals back to headquarters. But modern warfare could not be conducted in such a primitive manner. The scientists developed radar and the problem seemed solved: A directed beam of electromagnetic energy would detect any vehicle, anywhere—air, land, or sea. Of course, there was always the problem of discrimination. The blips on the radar screen would certainly provide enough warning, but warning of what? Perhaps the blip was a friend, not a foe.

Today, we have newer, more sophisticated sensing instruments. With the use of optical devices with extremely high resolution, distances of a few feet can be resolved from "eye-in-the-sky" satellites at altitudes of over 300 miles.

The human body also is being explored with new sensitive instrumentation. Today, physicians can see into the remotest corners of our bodies, using a wide variety of sensing techniques. Friend and foe stand out in sharp relief.

Probably the most far-out technique utilizes antimatter to obtain a "signature" of the metabolism that takes place in our brains. A radiopharmaceutical—that is, a nutrient such as glucose or a fatty acid—is injected into the bloodstream. This radioactive nutrient typically contains carbon 11, a form of ordinary carbon that has undergone cyclatron bombardment, making it radioactive. (A cyclatron is a machine that makes high-energy particles.) Carbon 11, however, produces a strange form of radiation. It makes positrons—the particles of antimatter that are antielectrons. Electrons exist in every atom of the body, so when a positron is emitted, it quickly meets up with an electron in its vicinity. When that happens, the electron and positron interact, and, lo and behold, simply vanish! In so doing they produce high-energy gamma rays. These rays are detected by instruments surrounding the organ that is absorbing the nutrient.

In this manner, pictures of parts of the brains can be obtained. This process is called positron emission tomography, or PET scans. There are also CAT scans, or computerized axial tomography, using X rays to invade the body.

These techniques, however, are not used indiscriminately; they involve, after all, high levels of ionizing radiation. Both X rays and gamma rays can and will damage surrounding tissue. Yet they are medically useful, allowing physicians to peer into the body while it is in the process of living. Any abnormalities will show up. In fact, these scans have also taught us much about normal processes. For example, differences in glucose takeup in the right side of the brain show up between musicians and nonmusicians when listening to music, indicating that musicians hear music differently from nonmusicians.

Safer kinds of sensing techniques that don't destroy tissue also are being employed, including magnetic resonance and ultrasound imaging, processes that I will explain in more detail in the following chapters.

In order to use such sophisticated sensing instruments, human beings themselves must also be sensitive instruments. All sensing is based on our body's electricity and on our ability to convert incoming signals—whether they are sounds, lights, touches, tastes, or smells—into electrical pulses. Indeed, the nervous system is the sentry system of the body. But just how a signal can be interpreted as a sound rather than as an image or a taste remains a mystery. In this part of the book I hope to show how quantum physics may help us to better understand this mystery.

CHAPTER 15

Singing the Body Electric

According to ancient Aristotelian philosophy, the world is made of four elements: earth, air, fire, and water. If we take earth to mean ordinary material or solid matter, then only fire requires further explanation in terms of modern physics. Today, we know that fire is an electronic affair. The ordinary fires that we see on our gas stoves or when lightning flashes in the upper atmosphere are visual because electrons are moving in quantum jumps from one atom or molecule to another. In so doing they emit the light we see. That romantic feeling that overtakes us when sitting by the fireplace and the fire itself are both created by electrons' quantum jumping. External quantum jumps produce the light; similarly, internal quantum jumps excite the nervous system.

How the Body Is Fire and Water

Inside our bodies a similar fire burns. Actually, burning in the old physics meant the chemical reaction of elements with oxygen, a reaction called oxidation. According to the new physics, the definition of oxidation has been generalized. When an element is oxidized, according to the old physics, it loses an electron to oxygen only; with new physics, oxidation means losing an electron to any electron acceptor. In so doing the substance is said to be *ionized.*

It is this "electron jump" that is observed as fire. Fires and all oxidation processes are always described by the simple process:

substance → ionized substance + 1 electron

The ionized substance, in losing an electron, is oxidized or burned. Thus, when hydrogen is burned, it becomes two ionized hydrogen atoms, called protons or hydrogen ions, and two negatively charged particles, called electrons:

molecular hydrogen → 2 protons + 2 electrons

When oxygen is present, it is attracted to the two protons and is able to accept the electrons. This takes place in two nearly simultaneous steps. First, a single oxygen atom absorbs the two electrons (the ones freed by burning hydrogen), becoming a doubly negative-charged oxygen ion:

atomic oxygen + 2 electrons → a doubly negative oxygen ion

Next, the oxygen ion attracts the protons and makes water, good old H_2O:

oxygen ion + 2 protons → water molecule

The two positively charged protons are attracted to the doubly negative-charged oxygen. This ability to attract electrons to its bosom is the reason why oxygen is such a good oxidizer. And that also is why oxygen is breathed: It has a high affinity for electrons, making each oxygen atom an ideal landing site for an electron. In other words, electrons in the body are attracted to oxygen atoms. It is the electrons that are the key to the reason we breathe: They carry the energy necessary for the body's vitality.

Even the water in the body is always on fire. That's because water is made of electrical molecules, each designed in such a way that it continually creates a tiny electric field around itself, called its *electrical dipole moment*. When other water molecules are in its presence, they tend to be pulled apart by the electric field produced by the dipole moment because they each have a tiny electric field themselves. An electrical dipole consists of two oppositely charged parts nearly on top of each other. The electric field produced by a neighboring water dipole thus pulls on the positively charged part, and pushes on the part that is negatively charged. So water molecules turn into protons as well as negatively charged oxygen-hydrogen molecules, called hydroxyls. In effect, water *burns* hydrogen:

water molecule → negative hydroxyl ion + a proton

At normal temperatures and pressures, the water dipole moment does not have quite enough strength to completely pull two protons from nearby water molecules. However, just one will do. The amounts of protons and hydroxyls found in ordinary water at room temperature normally are quite small. The number of free protons in water depends on the temperature of the water: Higher temperatures produce more protons. At any given temperature, however, there is an equilibrium concentration of protons. At equilibrium, protons and hydroxyls attract each other and form

108

ordinary water as quickly as they come apart to form ions again. A sinkful of water is a busier place than we might imagine!

The net proton concentration actually fluctuates in time but centers on an equilibrium value. The proton concentration found in water is a measure of the acidity of the water. This is called the pH factor, which stands for the *p*otential of *H*ydrogen present in a solution. The pH is measured by determining the number of protons (how many grams of protons) present per liter of water. Water contains few protons at normal temperatures. Its pH factor is 7. This number means that in one liter of water there are 10^{-7} grams of protons. A pH higher than 7 means fewer protons (note the negative exponent), while a pH lower than 7 means a higher concentration of protons. A pH of 7 also means an equal concentration of negatively charged hydroxyl ions. Water is taken as the standard in acidity; it is considered neutral, being neither acidic nor alkaline.

Solution with pHs lower than 7 are called *acids*. Solutions with pHs higher than 7 are called *bases*. A base solution acts to neutralize one that is acidic because it can absorb protons and bind them so that they are no longer free. A common human experience with acids and bases concerns indigestion, when your stomach is overloaded with food and producing HCl, hydrochloric acid. By taking in sodium bicarbonate, you turn an acidic stomach into a basic one. The chemical reaction goes:

$HCl + NaHCO_3 \rightarrow H_2CO_3$ (the fizz in seltzer water) $+ NACl$ (salt)
$\rightarrow H_2O$ (water) $+ CO_2$ (burp)

When carbon dioxide is injected into water under high pressure, the carbon dioxide tends to attract a hydroxyl and a proton, forming a unit called carbonic acid. Carbonic acid makes you burp when it turns back into carbon dioxide gas and water once again. Keeping a solution of carbonic acid cold and under pressure makes for good soda pop and champagne; uncorked, the bubbles you see are the inverse fires of protons reaching out and grabbing electrons from neighboring carboxyl units that are busy forming carbon dioxide and water all over again.

Electric Fields and How They Work

Without the fire and water inside our cells, we would not be able to think or move. Thinking and moving require the presence of electric fields in our nerve cells—literally, tiny electrical storms of fire.

The key factor in the body electric is water, especially insofar as it creates electrical fires involving other substances that dissolve in it. Water has a tendency to ionize certain substances, particularly acids—including the critical amino acids. Another substance readily absorbed or dissolved in water is salt, also a key element in the nervous reactions and actions of life. Water's ability to separate electrical charges, plus from minus, is what makes it such a miraculous substance. When charges separate, electric fields appear between them; such fields are force fields in space. They are visualized as lines of force, beginning on positive charges and ending on negative charges.

A force field is a condition of space that will act upon matter when matter is present in that space. The space of the body, filled by more than 65 percent water (the brain is 85 percent water), is no exception. Thus, where there's water there's fire, and that means electric fields and charges of electricity. It is the movement of electrical charges, under the actions of electric fields, that enables the body to communicate with itself. It is their ritual dance of fire that causes nerve cells to work in the ways they do.

Inside the Electrical Nerve

Body movements, both voluntary and involuntary, require the electricity of the nervous system. In fact, *every* living cell uses electricity in some manner or another. Muscle cells, for example, are elongated or contracted by electric fields. Luigi Galvani made the first observation of body electricity when he found that he could contract a dead frog's leg muscle by applying electrodes to the ends of the muscle. From this simple experiment, carried out in 1786, grew a whole field of inquiry called neurophysiology.

It also created quite a genre of horror films and stories. We see the mad Dr. Frankenstein bent over his "corpse" composed of dead parts assembled from the local graveyards. Eager Igor slobbers as he watches the doctor attach electrodes to the monster on the operating table; sparks run up the wires as Igor throws the switch; and . . . the monster breathes! Electricity has brought the dead being to life!

The experiments of Galvani and others not only inspired gothic fantasies, they brought to light the close connection of life and electricity. We now know that electricity controls the operation of nerves, muscles, and organs; that essentially, all of the body's functions are electrical in some way. The brain and nervous system generate both electric fields (called electrical potentials) and electrical currents (movements of electrons and positively charged ions). These currents also are detectable, measurable by sensitive electronic equipment outside the body. When an instrument detects an electrical current, it usually picks up on the magnetic field produced by that current; when detecting electrical potential, it measures the presence of an electric field.

The electrical potentials of nerve transmission have been measured directly. Measurements performed on electrical potentials in the muscles are called *electromyograms*, or EMGs; performed on the heart, they are called *electrocardiograms* (EKGs, or sometimes ECGs); on the brain, *electroencephalograms*, or EEGs. It is also possible to measure electric fields in the retina of the eye—an *electroretinogram*, or ERG.

The measurement of electrical potential within nerve cells enables scientists to determine just how a nerve cell conducts electricity as signals from one place in the body to another. A positive potential means that a nerve cell tends to repel any small positive electrical charge that is introduced, while a negative potential tends to attract the same test charge. When the potential is observed to oscillate with a particular frequency, it usually implies the presence of an electrical wave or alternating current.

To see how this works inside a neuron requires an examination of the neuron's electrical and physical structure. Neurons are present in rich abundance in the brain; in fact, it is thought currently that the human brain contains over 100 billion neurons—about the same number as the number of stars in our galaxy. No two neurons are ever exactly alike; their forms, however, are similar in character. There are three identifiable regions of any neuron: the cell body, the dendrites, and the axon.

The Cell Body

The cell body contains the nucleus and other biochemical structures, such as mitochondria, necessary to keep the living nerve cell alive. Usually the cell body is spherical in shape, although sometimes it is shaped like a pyramid.

The Dendrites

The dendrites are delicate, tubelike structures extending from the cell body that tend to branch and form bushy trees. Because of their length and small volume, they have large surface-area-to-volume ratios that provide plenty of opportunity for other neurons and muscles to pick up incoming electrical signals (that is, variations in electrical potential). In essence, the dendrites are like receiving antennae.

111

THE AXON

The axon extends away from the cell body and provides the pathway for nerve propagation. It is the pipeline by which an electrical message is sent from one nerve cell to another. The message originating at the cell body travels a long distance down the axon to other parts of the brain or to other nerve cells that pick up the message through their dendrites. Most axons are longer and thinner than dendrites and exhibit a different branching pattern. The dendrites tend to branch and cluster near the cell body, while the axon tends to branch and form its cluster at the end of the axon fiber, away from the cell body. In this way, the axon can communicate with other nerve cells.

Nerve Cell Communication

The nerve cell sends messages along its axon, and receives messages along its dendrites. In order for two nerve cells to communicate, they need to almost touch, but not quite. Their touching regions are actually separated by tiny gaps called synapses. Synapses are amazingly thin: typically, only about 9 billionths of a meter thick, or approximately 100 atomic diameters. (An atom is only around 1/100th of this length in diameter.) Molecules emitted within the synapse, called neurotransmitters, are much larger than atoms. Thus, with the emission of just a few, the synapse actually becomes crowded.

Each axon is covered by a thin membrane of cells called Schwann cells, comprising what is called the myelin sheath, an insulator for the nerve axon. This sheath is interrupted every few millimeters along the length of the axon by narrow gaps, apparent breakdowns in the insulation, called nodes of Ranvier. When a nerve impulse travels down the axon, it must jump the gaps, that is, hurdle the Ranvier nodes. It is an invigorating exercise. Nerve impulses decrease in strength as they shoot down the axon, but whenever they hit a node they are restored to their original strength. The nodes, therefore, act as tiny booster stations.

If, for any reason, the myelin sheath breaks down, the nerve cell becomes exposed. This breakdown occurs if the sheath becomes inflamed and swollen. When this takes place in a number of different tracks within the nervous system, you have a serious disease called multiple sclerosis (MS). There is some evidence that MS is caused by a viral infection, but it also may be due to a deficiency or abnormality in the fatty substance that makes up the myelin. Consequently, a breakdown in the sheath causes massive miscommunication: Messages do not get to where they are being sent. Typical symptoms include tingling and numbness, or weakness that may affect only one spot, one limb, or one side of the body. Other symptoms include general physical unsteadiness, temporary blurring of vision, slurring of speech, and even urinary incontinence. Surprisingly, in most cases all symptoms disappear after the first episode, and sometimes there are no further problems. But for some, unfortunately, symptoms recur. There appears to be no cure for MS.

How an Electrical Membrane Works

The neural axon membrane, like the outer membrane of all cells, is quite thin, only five nanometers in thickness. (A nanometer is a billionth of a meter; an atom is about .1 nanometer thick.) A typical membrane consists of two layers of fat, or lipid molecules, arranged as shown here in Figure 1.

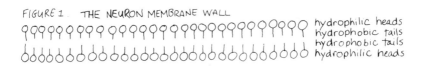

FIGURE 1. THE NEURON MEMBRANE WALL

hydrophilic heads
hydrophobic tails
hydrophobic tails
hydrophilic heads

The basic framework of all membranes is a double layer of lipid molecules. These molecules all share a common property: One end of the molecule attracts water molecules and is soluble in water, while the other end contains a hydrocarbon that makes it oily and water-insoluble. The heads of the lipids attract water and are called hydrophilic; the tails, formed from long chains of oily hydrocarbons, are called hydrophobic. When these molecules are introduced into a watery medium, the individual molecules spontaneously arrange themselves into a double layer, aligning themselves with their long tails perpendicular to their heads and with all their long oily chains pointing inward toward the middle of the double layer.

The double-layer structure of the membrane separates the waters of the body into inside-the-cell waters and those outside the cell. Embedded in the bilayer structure of the cell are complex proteins (Figure 2).

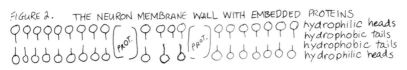

FIGURE 2. THE NEURON MEMBRANE WALL WITH EMBEDDED PROTEINS

hydrophilic heads
hydrophobic tails
hydrophobic tails
hydrophilic heads

The embedded proteins, which fall into five classes, are responsible for the neuron's ability to exchange electrical charges with its environment—a necessary step if the neuron is to function as an electrical telegraph wire. The classes are: ion pumps, channels, receptors, enzymes, and structural proteins. The pumps use the energy of metabolism to move ions and other molecules against concentration gradients, tending to maintain the ions and molecules in equilibrium. The channels are like gates for ions; without them, the bilayer would not allow any ions to pass through (that is why the lipids are a membrane in the first place). Each channel is "ionic specific," allowing certain ions to pass but not others. Receptors provide binding sites for specific molecules needed in the transmission of the neuron. Enzymes are placed in the membrane to speed up certain chemical reactions occurring at the membrane. Finally, the structural proteins link interconnecting nerves with organs and help in the formation of subneuronal structure. These embedded proteins are quite versatile and can provide more than one service. A single one could belong to all of the above classes or to any selected few of them.

How a Nerve Cell Propagates Information

In Figure 3 below, the ion concentrations inside and outside the axon wall are shown schematically.

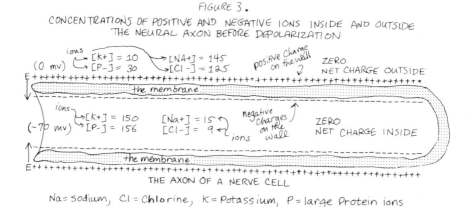

FIGURE 3.

CONCENTRATIONS OF POSITIVE AND NEGATIVE IONS INSIDE AND OUTSIDE THE NEURAL AXON BEFORE DEPOLARIZATION

THE AXON OF A NERVE CELL

Na = sodium, Cl = Chlorine, K = Potassium, P = large Protein ions

This figure shows the ion concentration state before the axon undergoes depolarization—the action wherein ions are exchanged across the membrane boundary, resulting in the initiation of an electrical pulse along the axon length. The concentrations of sodium (Na) and potassium (K), positively charged ions, are quite different on each side of the membrane. Similarly, the concentrations of negative ions of chlorine (Cl) and negatively charged large protein molecules (P) also are different on each side of the boundary. Although there are a larger number of both positive and negative ions inside the cell than outside, the net charge inside and outside is zero. Yet there is an electric field across the boundary!

MEMBRANE PHYSICS: ELECTRIC FIELDS WITH ZERO CHARGE

As a student of physics is taught, electric fields arise when there is an imbalance of charge; that is, a greater number of positive charges over negative ones in a region of space. Here no such imbalance exists but an electric field exists nevertheless. Why? Because of the membrane's selective ability to allow certain ions to pass through while restricting the passage of others. The ions on each side are simply not able to establish equilibrium concentrations, as they would if the membrane were fully permeable. The disequilibrium produces the electric field.

Suppose that sodium ions are trapped outside the cell and that potassium ions are free to pass across the membrane. Also assume that negative chlorine ions can pass across the boundary freely but that the larger negatively charged proteins cannot. From this model we can see how the inside of the neuron is at a lower voltage than the outside, like a battery cell.

The greater abundance of sodium ions outside the cell rush to the wall in an attempt to balance out the lack of sodium ions inside. The greater abundance of negative large protein molecules inside the cell also rush to the membrane and try to get out. The net result is an electric field across the membrane, with the inside appearing about 70 millivolts negative (a millivolt is 1/1,000th of a volt). The freely moving potassium and chlorine ions will establish concentrations that are in equilibrium with the electric field produced by the unequal concentrations of trapped ions.

Now, it turns out that potassium is not totally shut in while the nerve is resting (called the resting stage). Potassium and sodium ions both leak across the membrane because of the random nature of the protein gates—sometimes they are open, sometimes not. Consequently, the protein pumps tend to use up metabolic energy, maintaining the neural integrity. Each sodium pump can harness the energy of adenosine triphosphate (ATP) to exchange three sodium ions on the inside for two potassium ions on the outside.

BEING NERVOUS

In this manner, we have a potentially nervous situation: a kind of storage-battery effect, ready to strike sparks. When a stimulus is applied to the nerve cell body, the protein molecules embedded in the axon nearest the body undergo a conformational change, enabling sodium ions to get through and rush into the attracting negative voltage inside the axon. This causes the inside of the cell to momentarily become positively charged (remember that in the resting stage before depolarization the inside has a zero net charge), illustrated here in Figure 4.

FIGURE 4.

HOW AN ACTION POTENTIAL PROPAGATES DOWN THE LINE OF AN AXON.

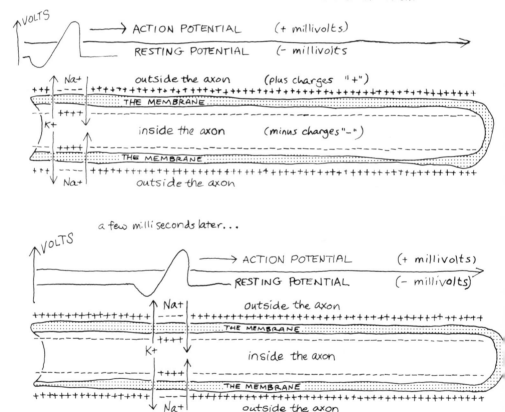

Now we have a chain reaction occurring in the potentially explosive axon. Potassium ions rush out, restoring the negative potential inside the axon. Meanwhile, the glut of positive charge stimulates the influx of more sodium ions, and the resulting "action potential" propagates down the line.

All of these processes, occurring at the tiny level of individual molecules undergoing minuscule quantum jumps of energy, result in the singing body electric. Your electrical system, consisting as it does of molecules on fire, enables you to be aware of your unique self. No other person conducts electricity through the body exactly the same way as you, yet everyone relies on the properties of the quantum processes of nerve-firing simply to know and to feel the experience of being a living human being.

CHAPTER 16

The Body Magnetic

Accompanying any electric field that is changing over time is a magnetic field. This fact has been known since 1864. Since the body electric produces electric fields that change over time—as, for example, whenever a nerve conducts a pulse along its axon— magnetic fields accompany them. One might say we all have magnetic personalities!

Until recently, however, no one had ever seen a magnetic field produced by the body's natural electricity. Such was the situation until a remarkable new technique was devised to detect these small magnetic fields. It was based on the 1964 discovery of Dr. Brian Josephson, Nobel laureate and physicist at the University of Cambridge, England. Josephson's junction (a joining of two surfaces with a microscopic gap between them), as it came to be known, enabled magnetic detection to be obtained many thousands of times more sensitively than ever before.

Physicians Samuel Williamson, Lloyd Kaufman, and Douglas Brenner of New York University were able to use a Josephson junction in 1975 to successfully detect the magnetic field produced by a human head while a person was thinking! (Becker, p. 240.) In fact, they found that the magnetoencephalogram (MEG) was a better, more accurate indicator of changes in the brain's activity than the electroencephalogram because the magnetic field is able to pass right through the dura, skull, and scalp without being diffused. The MEG locates the electric currents in the brain more accurately than an EEG.

This new technique is called nuclear magnetic resonance, or NMR. With it, pictures of the body at the cellular, and, perhaps, even the molecular level, may be obtained. And best of all, the technique does not harm the body at all. Since 1975, MEGs have been used with increasing success, particularly with the rapid growth in computer technology, as computers are needed to carry out the data analysis from a patient's MEG.

It is the ability of a magnetic field to pass out of the body, or

118

through it, that makes it so useful for sensing what takes place inside the body. In this chapter I hope to show how magnetism, coupled with quantum physics, may lead to a great breakthrough in our ability to sense the inner workings of the body human.

How Magnetism Works

We are all familiar with the common bar magnet, and have marveled at its seemingly magical ability to pick up and influence bits of iron, nails, and other magnets in its vicinity. The key to a magnet's ability to move metallic objects of iron, nickel, or cobalt, or to influence such other metals as copper, gold, and silver, is its magnetic field. Magnetic fields are produced from electrically charged matter whenever that matter is moving.

Our planet produces a magnetic field large enough to influence high-energy particles that impinge on us from the sun and other cosmic sources. It causes these tiny particles, consisting of electrons (negatively charged constituents of the atom), protons (positively charged nuclei of hydrogen atoms), and other exotic particles, to follow paths which are bent or spiraling along the lines of the magnetic field.

Indeed, many of us have heard that magnetic fields can be used to bend and focus streams of electrically charged particles as they move through space. The field of plasma physics was created just a few years ago in the hope that large enough magnetic fields could be produced to "squeeze" positively charged nuclei of hydrogen atoms together by altering their motions in a confined space called a *magnetic bottle*. In this manner the nuclei would be made to interact and produce radiant energy, which could then be converted to electricity.

Magnetic fields are measured in terms of their bending and focusing abilities. The stronger the field, the greater its ability to cause a moving electrically charged particle to deviate in its path of motion from a straight line. Magnetic fields are measured in units called *teslas*; the earth, for example, produces a field of about 50 millionths of a tesla. You observe the strength of this tiny field every time you note the movement of a compass needle.

The Body Magnetic

Your body's magnetic fields are produced from the motions of electrical charges within the heart as well as the brain. To observe these tiny magnetic signals, physicists use a *magnetocardiagram*

(MCG) for the heart's field, and a *magnetoencephalogram* for the brain's.

The MCG and the MEG provide information that can't be measured by looking at the body's electric field alone. The electric fields measured are produced by the body as the result of changes in the body's electrical-charge distribution. A *steady* flow of electrical charge will not be detected by instruments sensitive to such electric fields. However, the magnetic field that accompanies such a steady electrical charge is measurable with the MCG. This information often is useful when muscle or nerve tissue has been injured. (Cameron, p. 212.) When such an injury occurs, a steady current of electricity usually arises in the area, called the *injury current*. By detecting this current magnetically, it is possible to pinpoint the location of injury with great accuracy. Also, since injury currents exist in the heart before the onset of heart attack, early detection could be a life saver.

Magnetism and Blood Flow

Human blood is a veritable river of substances that are detectable electromagnetically. Blood contains moving electrical charges, and if the flow of those charges is placed in a magnetic field, the flow will be altered. This alteration produces a measurable electric field crosswise to the flow. Using an applied magnetic field and carefully isolating a blood vessel by surgical technique, a blood-flow meter can determine the velocity of blood movement by measuring this electric field. The greater the blood's flow velocity, the larger the electric field produced. From this information it is possible to determine if an organ is receiving a proper amount of blood in order to sustain itself.

Magnetic Resonance, a Window into the Molecular Body

The most important use of magnetism in the body comes from magnetic resonance. Although most of us think of our bodies as opaque, there are actually two windows in the electromagnetic spectrum through which we can "see" inside it. These windows are frequency regions within the spectrum of electromagnetic energy, frequencies outside those we normally see with our eyes.

Our eyes have adapted to the central frequencies of the sun. We call these frequencies ordinary light, and their wavelengths range from 400 to about 700 nanometers (a nanometer being a billionth of a meter). However, the total range of the electromagnetic spectrum is quite vast, ranging from gamma rays with wavelengths as small as 10^{-16} meters (so infinitesimal that the diameter of an atom is around 1 million times longer), all the way to radio waves with wavelengths 1 million meters (625 miles) in length.

Within that range exist two windows. The first, quite well known to most of us, is the X-ray window, with wavelengths ranging from 10^{-10} meters to 10^{-12} meters. Through our knowledge of quantum physics, we know that for every wavelength there is an associated quantum of energy: the shorter the wavelength, the higher the energy of the associated quantum. All electromagnetic radiation consists of quanta, called photons. Thus, gamma radiation consists of extremely high-energy photons of about 1 billion electron-volts (eV). (One electron-volt is the energy required to move an electron through 1 volt of electric potential.) To grasp the significance of this number, consider that it takes only about 14 eV to pull an electron away from an atom of hydrogen.

X rays possess a range of energies from 1,000 to 1,000,000 eV. A large dosage of X rays applied to the body will case damage to the cellular structure and alter the genetic code because X rays penetrate the skin with little or no absorption. Higher energy radiation, in turn, will cause more damage. But when the energy of the photon is smaller than about 1,000 eV, the skin is able to protect the internal organs of the body through absorption. When the energy is reduced to a range from 100 eV to 1, corresponding to the range from ultraviolet to visible light, the skin acts as a screen. Ultraviolet light can act on melanin molecules in the skin, causing sunburn. Ordinary light that we see with our eyes simply reflects from the skin.

With a reduction in photon energy, corresponding to lengthening wavelength, the body becomes more and more opaque. Infrared rays, at about 800 nanometers, and microwaves, at about 1 meter, are not able to penetrate the body.

However, with a further lengthening of wavelengths, corresponding to lower energy photons, a second window begins to appear in the electromagnetic spectrum. This window corresponds to extremely low-energy photons, ranging from 10^{-7} to 10^{-11} eV, with wavelengths from 10 meters all the way up to 100 kilometers. This range corresponds to radio waves: UHF, VHF, shortwave, standard broadcast, and ionospheric waves. This latter window is called the NMR window—the safest window we can use to peer into the body using electromagnetic waves. The body is actually transparent to these waves, whose energy is so small that they cause no tissue damage whatsoever.

The body's dimensions are so small in comparison to these wavelengths that, under normal circumstances, little wave scattering or absorption occurs. However, if the body is placed in a magnetic field, the whole situation changes and these waves will be selectively absorbed.

NMR—Seeing into the Body by Resonating with Protons

To grasp how these low-energy electromagnetic waves can be used to peer into hidden organs of the body, it is necessary to look at how our bodies can be induced to manufacture them. The trick is to find a way to excite the tiniest parts of our bodies; in particular, the nuclei of hydrogen in every drop of water in every cell of our bodies. These nuclei are protons—tiny particles which make up not only the centers of hydrogen atoms but also every atomic nucleus.

As any eighth-grader knows, magnets produce force fields in the space surrounding them. This field is called a magnetic dipole field—"dipole" because the fields are generated from two opposite poles, as, for example, the north and south pole of a bar magnet. The earth's magnetic field also exhibits this polarity, as, indeed, do all magnetic materials.

As we have noted, magnetic fields also are produced by electric currents. Whenever an electrically charged particle of matter moves, it makes an electric current and a magnetic field. When a charge moves in a circle or spins about an axis, it produces a magnetic dipole field.

Making up every atom are both electrons and protons. Each particle is electrically charged and each can be visualized as a spinning ball, whirling about itself the way the earth turns on its axis. A rotating electrical charge is no exception to the moving electrical charge rule: It produces its own magnetic dipole field.

Again, of specific interest to us here are the tiny magnetic fields produced by the proton-nuclei of hydrogen atoms. These magnetic dipole fields produced by the protons (and, for that matter, by all electrically charged particles that spin) have a special name of their own: *magnetic moments*. No, magnetic moments aren't those romantic interludes when our hearts go pitter-patter; the word "moment" refers to the turning movement of any body spinning or rotating about an axis. A flywheel, for example, has a moment of inertia. By "inertia" I mean the measure of its ability to continue rotating with the same rotational speed whenever the flywheel is set in motion. A proton also has a moment of inertia, and because it is electrically charged, it has a significant magnetic moment as well.

As a schoolchild quickly learns, when a tiny magnetic moment

122

is put in a magnetic field it tends to align itself so that it points along the direction of the magnetic field. A familiar example is seen in the compass, when the tiny magnetic moment of the magnetized compass needle points in the direction of the earth's magnetic field.

However, these alignments do not take place instantly. Because the magnetic moment is caused by the spinning charge, it behaves very much like a gyroscope or spinning top; see Figure 5. When the top is released and allowed to hit the floor, it tends to move in a certain manner as it tries to align with the earth's gravitational field. Without this spin, the top would simply topple over; with it, it moves so that the spin axis precesses or wobbles.

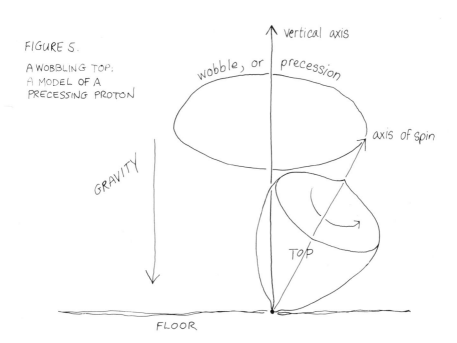

FIGURE 5.

A WOBBLING TOP:
A MODEL OF A
PRECESSING PROTON

Vertical axis

wobble, or precession

axis of spin

GRAVITY

TOP

FLOOR

To see how the top precesses, take a pencil and hold it at the tip with one hand and at the eraser end with the other. Imagine the pencil to be an axis running through the top. If you hold the eraser end below the lead end and tip the point toward yourself while keeping the eraser end fixed, you will be able to rotate the point in a circle while keeping the eraser end still. This is how a spinning top will turn when it hits the floor.

A magnetic moment placed in a magnetic field exhibits a similar precession. Anything moving in a circle has a frequency, and the movement of the magnetic moment is no exception. The frequency of rotation depends on the strength of the magnetic field. The stronger the field, the higher the frequency.

THE QUANTUM PHYSICS PICTURE OF NMR

According to quantum physics, protons placed in a magnetic field will behave in a manner analogous to the description given above. I say analogous because in quantum physics the picture of precession needs refinement. Instead of precessing moments, each proton will either flip over or not flip over when placed in a magnetic field. The proton's quantum behavior allows its magnetic moment to take on only one of two possible directions relative to the direction of the magnetic field. These directions either are parallel to the field (up) or antiparallel to the field (down). It turns out that down protons have higher energy than up protons. The energy difference is equal to the simple product of a constant of proportionality, called Planck's constant, times a frequency. And this frequency is just right to make radio waves.

It is therefore possible to excite protons placed in a magnetic field by aiming radio waves at them with just that particular frequency. The frequencies that excite protons are those that shine through the lower electromagnetic window. Of course, without the presence of the magnetic field, the protons would simply ignore these radio waves. But in a magnetic field, the protons actually can absorb radio waves like eager little radio receivers. These proton receivers are also able to broadcast similar radio signals to receivers outside the body. This is a happy circumstance: It makes it possible to form images of bodily structures by targeting the locations of these broadcasting protons within the body.

To grasp this concept more clearly, imagine placing yourself in a magnetic field. The protons in your cells will immediately begin to precess about the direction of the field. Most of those protons will point upward, with a smaller percentage pointing down.

When the radio wave is turned on, the protons gain energy, causing some of the up protons to flip into down protons. But since the precession frequency of the spinning protons depends on how strong the magnetic field is, only those protons that have the same frequency as the radio wave will actually flip. To make this clearer, suppose that the magnetic field strength is not the same value everywhere in the body, due to both the different tissues and the shape of the magnet itself. These precessing protons that happen to be in the right magnetic field will flip over, absorbing the radio wave; protons in a different part of the body that are not in

124

same field strength will not flip because the radio frequency is not the same as their precession frequency.

In other words, some of the protons will be attuned to the radio waves coming into the body in much the same way that a radio receiver is tuned to a particular station. A radio in another room is not necessarily tuned to the same station. In your body, depending on where the protons are located, in which cell, in which water molecule, some will be tuned to the incoming radio-wave signal, some will not. By varying the frequency of the incoming radio wave, selective protons throughout your body will respond.

MAGNETIC SLICES OF THE BODY

As a result, it is possible to cause different parts of the body to selectively absorb radio-wave energy by placing the body in a spatially varying magnetic field. One way to accomplish this is by using a magnetic field with a built-in variability called a field gradient. For example, suppose you wanted to look at the heart while it was beating.

By varying the magnetic field in a field gradient, as shown in Figure 6, you would produce magnetic "slices" in the heart, like sliced bread, with each slice in a different magnetic field. Once the radio waves are turned on, they would excite only protons in the corresponding slice.

FIGURE 6.

MAGNETIC SLICES OF THE HEART

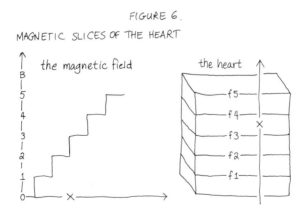

WHAT THE SLICES TELL US

Now that the protons have been excited, what good is it? What do excited protons tell us about the body? Well, first of all, as soon as the radio waves are turned off, the protons will try to return to equilibrium. That means that the down protons will try to quantum jump and become up protons with less energy. They thereby will emit radio waves at exactly the relief frequency that excited them in the first place. It is the character of these radio broadcasts that tells us something about the structure of the slice emitting them.

The strength of the emitted signal depends on how many protons there were in the slice; many protons produce strong signals, fewer produce weak ones. Thus the strength tells us something about the water density of the cellular tissue. As one moves across an organ, the density of water will change, and, therefore, the variation in intensity or strength of signal will exhibit that change, giving us a picture.

Cancer cells exhibit anomalous signals. As a consequence, NMR not only can show us more about tissue structure, it also may be the method for telling us where cancer microsites are occurring.

The chief reason for using NMR is to obtain a noninvasive image of the physiology and functioning of living cells in the body. An NMR image of the stomach could reveal that you are having trouble assimilating certain foodstuffs or specific minerals, such as phosphorus. Probably the most important use of NMR is the study of brain tumors, particularly in cases where local chemical effects are important to study.

NMR is a relatively new technique for peering into the body. Along with other techniques, such CAT and PET scans, X-ray absorption studies, and ultrasound, it is highly reliant on the use of computers to obtain images. With improved computer technology, NMR will eventually replace conventional X-ray imaging. I will describe the process of ultrasound imaging in the next chapter.

The major difficulty with NMR is that it takes a long time to obtain an image. While it takes only two seconds to obtain an X-ray, an NMR takes a number of minutes. While NMR signals are much weaker than X ray, X rays cannot even tell us the difference between a living body and a cadaver. Indeed, X rays all look alike. NMR scans are entirely different, using much smaller quantum units of energy. Too, NMR appears to be the most benign method of obtaining detailed information about cellular functioning.

CHAPTER 17

Sounds and Ears

Sound is not only heard but felt. It is not only used to carry information to the ear, but, thanks to diagnostic ultrasound scans, to the eye as well. Because sounds are waves, they have vibratory patterns, patterns that help scientists and physicians to actually map out regions of the body heretofore inaccessible to human inquiry.

Sounds We Hear and Don't Hear

Sound provides probably the most important kinaesthetic sense of human awareness. Just imagine a day at the beach without the slap of the waves, a meal without the crunch of food or the solace of conversation, a night at the theater without audible dialogue or the thunder of applause. Music probably lifts the spirits more than any other sound—except, perhaps, the sound of your sweetheart whispering sweet nothings in your ear. Sounds sway us and, unfortunately, swat us, too. These days sound pollution is rampant: the pulse of loud radios, the wail of car alarms, the screech of subway wheels try the patience—and damage the eardrums—of beleaguered city dwellers. And yet, despite federal regulations that limit noise levels, sound pollution is a necessary evil of modern life.

Sound Physics and Psychophysics

And so we come to the physics of sound. Because sound is a wave, there are two forms for its measurement—intensity, and frequency (or wavelength)—humans being sensitive to both. First, let us look at intensity.

SOUND BALLOONS: A PICTURE OF INTENSITY

Intensity is measured in watts per meter, squared, or watts per unit area, the rationale being that a sound wave, initially emitted

127

from a small source, spreads out and fills the space around it, much like a balloon blowing up at the speed of sound. With each increase in the sound-balloon's radius, the area of its surface increases mathematically as the square of that radius. A sound balloon 1 meter in radius has an area of about 12.5 square meters, while one grown to 2 meters' radius has an area of 50 square meters. However, any given sound balloon carries a fixed amount of energy that spreads out over an ever larger area as the balloon grows larger. It is like trying to spread a tablespoon of peanut butter over many slices of bread—soon there isn't enough to go around. Thus, the sounds you hear become fainter as the sound source moves away from you.

Human sensitivity to sound intensity is amazing. Sound intensity is proportional to the square of the amplitude of the sound wave. Amplitude is a measure of the eardrum's displacement, when it vibrates sympathetically with the incoming sound wave. If one knows the sound intensity reaching the ear, therefore, one can determine the amplitude of eardrum vibration. The faintest sound intensity that the healthy ear is capable of detecting is 10^{-16} watts per square centimeter; this figure translates into an amplitude variation of approximately 10^{-9} centimeters—smaller than the diameter of a hydrogen atom!

It is hard to imagine that our ears are able to sense sounds with such small amplitudes. Eardrums, however, do not respond to sound waves as one might think. Suppose that the amplitude of a sound wave doubled; would you hear it twice as loud? The answer is no. First, your ability to sense sound depends on the intensity, which is the square of the amplitude. You therefore might think that you would hear it four times as loud, but, again, it is not so simple. The perception of loudness is not proportional to the intensity, either. The relationship between intensity and perceived loudness is best measured when the intensity is expressed in special units called decibels (dB)—a unit named after Alexander Graham Bell.

Imagine sitting in a quiet room on a bright fall day. The leaves on the trees in the backyard are rustling in the wind; you can hear them. When the wind stops, you perceive the sounds of your quiet room. But then, suddenly, a construction crew starts wrenching open the street in front of your study with an ear-splitting jackhammer.

Each of these sounds is perceived quite differently. The quiet room is perceived at a certain level, which we shall designate 1, which corresponds to a sound intensity of 40 decibels. The rustling leaves register at a level of .25, or a decibel level of 20, while the jackhammer is around 200, or 115 decibels.

In Figure 7, we see a graph relating loudness of sound, as perceived by humans, to its intensity. I developed this graph using techniques developed by S. S. Stevens, who discovered the relationship between the loudness perceived by humans and the sound intensity. Again, this relationship is understood best when the intensity is measured in decibels. One decibel corresponds to the sound made by the faintest movement of the eardrum possible—the sound of the eardrum vibrating through an amplitude less than the width of a hydrogen atom. A change in intensity of a factor of 10 produces a change of 10 dB, while a change in intensity of a factor of 100 produces a change of 20 dB.

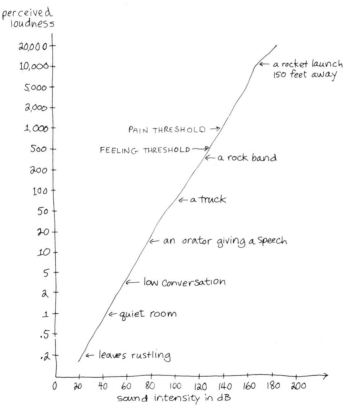

THRESHOLDS OF SOUNDS WE HEAR AND FEEL

FIGURE 7.

A simple linear increase in sound intensity, therefore, is not perceived as such. A sound intensity of 20 dB (leaves rustling), which corresponds to an intensity 100 times the minimum intensity perceivable by humans, has a loudness perception of .25. A sound intensity of 80 dB (an orator delivering a speech), corresponding to an intensity 10 billion times minimum intensity, is perceived as 15.8. And a sound intensity of 130 dB (a loud rock band), with an intensity 10 trillion times the minimum, is perceived as 501—near to the pain threshold.

A MEDIUM MEDIATES SOUND WAVES

Sound waves would not exist without a medium in which to propagate. The sounds that we hear are due to vibrations of air—the medium for the sound wave. Sound waves also propagate through water, the brain, muscle, fat, and bone. The energy carried by the sound wave depends on the properties of the medium, as well as the amplitude of the wave. Sound waves also travel at different speeds in different media because various media possess differing densities. Table 8 shows values for density and velocity for several media of interest found in the human body.

Table 8

Medium	Density kg/m^3	Velocity m/sec
Air	1.29	331
Water	1,000	1,480
Brain	1,020	1,530
Muscle	1,040	1,580
Fat	920	1,450
Bone	1,900	4,040

In air, sound travels with a speed of about 331 meters per second (1,085 feet per second, or .2 miles per second).

FREQUENCY IS VIBRATION IS THE SOUND OF MUSIC

The frequency of sound waves is the second measurable factor in the physics of sound. A sound wave is a mechanical disturbance of the medium; that is, it causes the medium to undergo a consecutive series of expansions and contractions due to the variability of its pressure. Thus, the air in the vicinity of a sound source is undergoing a series of compressions and rarefactions, or decompressions, wherein the air density under compression is increasing and, under rarefaction, decreasing. These variations are designated longitudinal because they take place along with and in the same direction as the sound wave travels.

The variability rate of these compressions and rarefactions is referred to as the frequency of the sound wave. If you were to place your ear at one location, toward the direction of the sound wave's travel, your eardrum would begin to vibrate at the frequency of the wave. The pressure variations cause the drum to move in and out sympathetically, in resonance with pressure changes. However, there is a limit to the eardrum's frequency sensitivity: Humans can hear frequencies ranging from 20 hertz (H_2) to about 20,000 hertz (1 hertz equals 1 cycle per second). With increasing age, upper-frequency sensitivity decreases. Figure 8 below shows how hearing sensitivity changes with age.

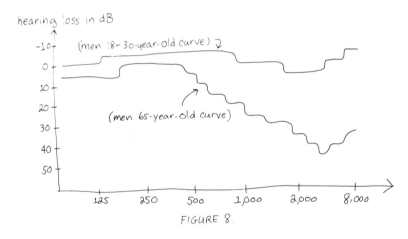

FIGURE 8.

The decrease in frequency sensitivity with age is sometimes due to a rather simple problem that is easily correctable—accumulation of earwax. It turns out that the older we get, the faster wax collects. Another possibility is the disease called otosclerosis, caused by an abnormal growth of spongy bone at the entrance to the inner ear. The disease immobilizes the base of the stirrup, a tiny bone through which sound waves must pass into the inner ear.

Natural deterioration of the hearing mechanism gradually causes hearing impairment, but not necessarily deafness, and for this there appears to be no cure.

131

Inside the Human Ear

The ear can be divided into three distinct regions, according to location and anatomical function. The *outer ear* is responsible for gathering in sound waves and funneling them to the eardrum. Moving inward, we come to the *middle ear*. This region starts at the eardrum, and it functions as a mechanical transformer. It includes the three smallest bones in the human body, the *ossicles*. The middle ear thereby transmits sound information to the *inner ear*, which consists of fluid-filled, spiral-shaped *cochlea* containing the *organ of Corti*, the primary auditory receptor. From here tiny hairs that line the cochlea convert the sound waves into electrical impulses that travel to the brain.

Recent work by physicists William Bialek and Allan Schweitzer indicate that the ear is capable of detecting sound at a level limited only by the principle of uncertainty in quantum physics. (Bialek, p. 1.) In other words, the inner ear performs as a quantum mechanical instrument. If correct, it opens up an entirely new possibility for the hearing mechanism. According to my view, this mechanism is strongly determined by what is known as the observer effect of quantum physics. Briefly put, to hear anything intelligently it is necessary to pay attention to what is being heard. For this to happen, cells in the ear must act coherently, much as atoms act inside a laser.

The key insight brought forward by such quantum physical considerations is that of feedback. The hairs in the organ of Corti must exist in a nonequilibratory configuration, meaning that normal temperature effects will not destroy their ability to act coherently. Together their vibrations send a coherent message to the brain, and the brain responds by sending back a feedback message to the hairs inside the cochlea. That feedback results in the actual perception of sound. Without feedback, even though sound waves would reach these hairs, no sound would be perceived.

THE OUTER EAR AND WHAT IT DOES

The outer ear is the external auditory canal that terminates at the eardrum, not the visible part of the ear that you hook your glasses around. That is called the external auricle, or pinna, and it actually can be removed without any appreciable loss of hearing. However, the pinna collects sound waves and funnels them into the middle ear.

Humans can often use the pinna's design principle to increase hearing by cupping the hand around the ear; by so doing, a 6 to 8 dB gain in intensity can be achieved.

However, it is the design of the ear canal that is chiefly responsible for the ear's sensitivity to sound. The canal is about 2.5 centimeters long (about 1 inch), with the diameter of a pencil (as some of you who have gotten pencils stuck in your ears know). The canal functions much like a tiny organ pipe closed at one end. As you may know, organ pipes resonate or vibrate sympathetically when certain notes are played; the familiar experience of blowing across an empty Coke bottle illustrates the principle involved. The ear canal is no exception to this principle, and it is the right size to resonate the sounds that are important in music and speech.

What determines which sounds will resonate in a given pipe? The length of the pipe. In order for a sound wave to resonate, the pipe must be long enough to hold at least one-fourth of the sound's wavelength. To grasp the concept of a wavelength, imagine a sound wave filling the room in which you sit. It begins at the source of sound and travels in waves or ripples of compression (air molecules crowding together) and rarefaction (molecules spreading apart) to all ears ready to receive them. The distance between any two successive compressions or between any two successive rarefactions is the wavelength, L. Since that wave is traveling through space at the velocity of sound, there will be a certain number of wavelengths passing into an ear each second. The higher the velocity, the greater the number of passing wavelengths. The sound velocity divided by the number of wavelengths passing into the ear canal is the number of vibrations per second entering the ear canal. This is called the sound frequency.

Your ear canal is 2.5 cm long. This means that it will resonate when a sound wave entering it has a wavelength of 10 cm. The velocity of sound is about 33,100 cm per second. A little arithmetic shows that the ear canal is designed to resonate at a frequency of 3,310 cycles per second, or 3,310 hertz.

Now of course, we hear other frequencies, too, so why is the canal designed for 3,310 hertz? Try blowing into the Coke bottle a little harder; you will observe that the bottle will resonate at a higher frequency. So too your ear canal resonates with higher frequencies—6,620 hertz, 9,939, even higher. Although lower frequencies will not resonate in the ear canal, their waves will still be able to reach the eardrum.

In other words, the resonance effect simply enhances sound within this range of frequencies around 3,310 hertz, but does not appreciably damp out frequencies lower than the resonant frequency.

The eardrum, or tympanic membrane, is about .1 millimeter thick, with an area of 65 square millimeters. When it vibrates, its movement is coupled to the tiny ossicles of the middle ear. The eardrum's movement is extremely small and quite complex in response to even the simplest sound patterns. At fairly low frequencies, below 2,400 hertz, the membrane vibrates pretty much as a whole. As the frequency increases past 3,100 Hz, the vibration pattern changes so that different parts of the drum vibrate independently.

But hearing a sound depends on more than just the frequency, it also depends on the loudness or amplitude of the sound wave. The smallest amplitude that can be heard is called the hearing threshold. Think of amplitude as the physical movement of the eardrum vibrating; this vibration ranges over a very small distance. The amplitude of the vibrations at the hearing threshold, with a frequency of 3,100 Hz, is only 3.5×10^{-10} cm, smaller than an atom of hydrogen. With an increase in intensity, there is a corresponding increase in the amplitude of eardrum vibration.

It is fascinating to compare the amplitude of vibration of the eardrum, produced by a range of ordinary sounds, with the minimum amplitude at the threshold of hearing. Suppose we label this ratio as β.

A whisper with a dB of 20 produces a β of 10; a low conversation at 60 dB produces a β of 1,000; a truck roaring past, 110 dB, produces a β of over 315,000. And a dB of 130, corresponding to the pain threshold of hearing, produces a β of over 3,000,000. However, even at the pain threshold, the eardrum is vibrating only through 1,106 millionths of a meter. At the low-frequency threshold of hearing (20 hertz), or 80 dB, the eardrum amplitude is 5.4 millionths of a meter. By way of comparison, the wavelength of visible violet light is .4 millionths of a meter.

Hearing thus involves very small amplitude vibrations of the eardrum, leaving Bialek and Schweitzer to argue that if it was not for quantum mechanical coherence, the simultaneous vibration of many cells together, existing on a time scale corresponding to the frequency of sound waves, we would not be able to hear anything at all. Hearing must involve quantum processes.

THE MIDDLE EAR: HOW NATURE OVERCOMES A MISMATCH

On the other side of the eardrum is the middle ear—an air-filled chamber containing three connecting bones, the ossicles. These bones act as mechanical "middlemen," transmitting the sound from the eardrum into the inner ear. The bones of the middle ear are needed because there is a mismatch between the sound entering the outer ear and the chamber of the inner ear; the middle ear overcomes this mismatch.

The inner ear actually consists of a water-filled chamber, the cochlea. The middle ear, therefore, somehow must convert or transform the sound that exists in air in the outer ear into sound that can exist in water in the inner ear. A device which transforms sound in this manner is called a transducer.

Before we look at the actual movements of these transducer bones, it will be useful to understand what happens to sound waves when they pass from air into water. The boundary between air and water is a surface, called the boundary surface.

Whenever sound waves reach a boundary surface, two things must happen: The waves must partially reflect and partially transmit across the boundary. The relative amounts of sound energy reflected/transmitted depend on the relative speeds of the sound in the reflecting/transmitting media, and respective media densities. Water is much denser than air, and the speed of sound in water is much higher than in air. This phenomenon produces a rather large mismatch in the ability to transmit sound from air into water. A little experience in a bathing pool will quickly convince you that this mismatch exists. When swimming underwater, it is nearly impossible to hear anyone shouting at you from above the surface. Similarly, if you try to make any sound under the water, little of it is heard in the air above. A measure of the ability to transmit energy from one medium into another is called the impedance (Figure 9).

FIGURE 9.

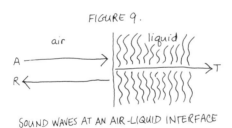

SOUND WAVES AT AN AIR-LIQUID INTERFACE

It is this huge mismatch in air/water impedances that makes the work of the middle ear so necessary; its job: to overcome that mismatch by providing a mechanism to amplify the signal transmitted to the inner ear. This mechanism is cleverly contained in the ossicles. They consist of three bones called the *malleus* (Latin for hammer; from it we get the word mallet), M; the *incus* (anvil—where else would a hammer pound?), I; and the *stapes* (for stirrups), S. Together the M-I-S system produces the necessary amplification through the mechanical laws of physics.

Here's how it works. First, M is connected to the eardrum;

135

consequently, when the drum vibrates, M moves back and forth. The effective cross-sectional area of M is the same as the eardrum area, around 65 mm². Next, M connects with I, which, in turn, connects with S. The stapes, S, are connected to an oval window, the entrance to the inner ear. It is through this window that sound information must pass in order to be heard. This window is part of the cochlea. Here, nature provides a sound boost by making the oval window smaller than the eardrum. The oval window has a cross-sectional area of only 3.2 mm².

The boost comes from the law of mechanical advantage, the same law you use when you pound a nail into a wall or when you ice skate or ski; the advantage results from spreading a force over an area. The same force applied over a large area has a smaller pressure than when applied over an area that is small. If a force of 10 pounds is applied to an area of 1 m², it produces a pressure of 10 pounds per m². If that same force is applied to an area of 1 mm², the pressure increases to 10 million pounds per m².

That is why an ice skater cuts a path in the ice, while someone wearing ordinary shoes leaves no mark. It is also why hitting a hammer against a wall leaves a dent, but hitting it against a nail drives the nail into the wall. You apply the same force over a smaller area, the area of the nail head, and thus achieve a larger pressure. The skier uses the law of mechanical advantage in the opposite manner (he or she uses a negative mechanical advantage). By spreading his weight over a larger area, the area of the skis, he exerts less pressure along the length of the skis, making it possible to maintain balance. A snowshoe also works using negative mechanical advantage, allowing you to walk across snow and not sink in (which would be the case with ordinary shoes).

The M-I-S system works in the same way: sound hitting the eardrum at minimum intensity corresponds to a small pressure on the membrane. That pressure, multiplied by the area of the eardrum, yields the force hitting the M-I-S. That force travels through th M-I-S system and reaches the oval window, where, when it is divided by the smaller area of the oval window, yields a larger pressure. The pressure increase on the oval window corresponds to an intensity increase of a factor of 400.

A second advantage derives from the way in which the M-I-S bones are connected. Together they act as a lever and fulcrum, which provides an additional intensity increase of about 29 dB. And so the middle ear performs its little miracle, nearly overcoming the 30-dB drop produced by the mismatch of the air/water impedance.

THE INNER EAR: CONVERTING SOUND TO ELECTRICITY

The inner ear, which consists of the small, spiral-shaped organ called the cochlea, is hidden deep within the bony skull. Hence, the cochlea is probably a human's most protected organ. It is filled with a clear fluid known as *perilymph*, which chemically is similar to the cerebrospinal fluid that bathes the brain and spinal cord, and physically is not very different from water. Thus, as we probe further, we find ourselves approaching a transition from the airy outside world to the fluid-filled electrical inside world of the brain and nervous system. The inner ear is the first step in that journey.

The oval window is the entry point to the cochlea, and the cochlea is actually a part of a larger structure called the bony labyrinth, which consists of the semicircular canals and cochlea together. The semicircular canals are three loops containing perilymph; they function not in hearing but in balance. Each loop lies in a plane, and the three planes are perpendicular to one another, like the walls and ceiling at the corner of a room. Thus, this "three-ring circus" is capable of analyzing motion in any direction in terms of the three dimensions of space. This enables each of us to keep our balance when we move.

A ride on a roller coaster will quickly convince you of the operations of the circus. You become dizzy after or during such a ride because the fluids in these canals do not have a chance to communicate with your brain just exactly in what direction your body is moving. Airsickness and seasickness also result from this miscommunication coming from the circus. The problem is that the fluids are circulating in all three rings, giving contradictory messages to the brain. Drugs, such as Dramamine, interrupt those signals so that you remain calm in spite of the dizzy situation in which you may find yourself. Movement of the head in the plane of the horizon causes swirling of the fluid in the horizontal ring. Fluid in the other loops does not swirl because they are perpendicular to the movement. A similar phenomenon occurs with movement in the other directions.

But back to the realm of hearing. It is the cochlea that is responsible for sending sound information to the brain by converting sound waves into electrical signals that are transmitted by the nerves. The cochlea is a spiral-shaped structure containing cells that act as auditory receptors. The movement of the oval window caused by the incoming sound wave produces pressure changes in the cochlear fluid. These pressure changes are picked up by the auditory receptors.

The cochlea actually consists of three separate compartments wound into a single spiral. It is important to realize that the whole bony labyrinth, including the cochlea, is actually a hollow space in bone structure. The sound properties of the cochlea do not depend on its spiral shape, which is merely nature's way of saving space. (If the cochlea were straightened out, it would stretch out to around 3 cm.)

The most important structure in the cochlea is called the *basilar membrane*. A sound wave entering the oval window produces a wavelike ripple in the basilar membrane, which vibrates according to the frequency of the incoming signal. Imbedded in the membrane are tiny hairlike cilia, extending from what are called hair cells; altogether, there are over 23,000 such cells in the cochlea.

When the basilar membrane vibrates, it causes the cilia to move laterally, much like a field of wheat rippling in the wind. Depending on which part of the basilar membrane is excited, this rippling generates a series of electrical pulses. A high-frequency wave near the end of the cochlea does not send a corresponding high-frequency electrical pulse along the nervous system. Instead, a *series* of pulses is sent from the excited region, indicating that the signal is high-frequency. If the frequency is lower than 1,000 hertz, the nerve-pulse frequency is the same as that of the sound wave.

Making Body Pictures with Sound

Sound waves, like all waves, come in a variety of wavelengths and frequencies. As stated earlier, although the human ear responds only to frequencies between 20 and 20,000 hertz, bats can hear and emit sound waves ranging from 30,000 to 100,000 hertz (30 to 100 kilohertz). With increasing frequency there is a corresponding decrease in wavelength, according to the frequency wavelength relation:

$$\text{wavelength} = \text{sound velocity/frequency}$$

When the frequency of sound reaches millions of hertz, or megahertz, the sound is called ultrasound, well beyond the province of the human ear. As I pointed out earlier, any sound wave will be partially reflected and partially transmitted when bridging one medium to another, and ultrasounds are no exception. However, because ultrasounds have such high frequencies, they also have short wavelengths; it is this property, coupled with their reflective abilities, that lets us use them to see inside the body.

Table 9 shows some typical ultrasound wavelengths and media encountered in making ultrasound pictures with a 1 megahertz sound wave.

Table 9

Medium	Frequency (MHz)	Wavelength (mm)
Air	1	0.331
Water	1	1.480
Brain	1	1.530
Muscle	1	1.580
Fat	1	1.450
Bone	1	4.040

The basic idea of making an ultrasound picture of the body is similar to the principle used by bats and by naval ships equipped with sonar. A signal is generated outside the body by a transducer, which acts as both a signal generator and receiver. The transducer is usually placed in close contact with the body surface. Water, or a jelly paste, is applied to the skin to eliminate the air interface between the transducer and the skin (this eliminates much reflection from the air/skin interface mismatch, which is similar to the air/water mismatch).

The generated sound waves are then transmitted through the body until they encounter the surface of an object, such as a fetus or organ, where, according to the description above, the signal is reflected due to the mismatch in impedance at the boundary of the object. The reflected signal takes some time to travel back to the transducer, a delay that yields a measure of the distance from the transducer to the object. By moving the transducer along the surface of the body, one receives a series of signals that can then be stored as a picture, producing an outline of the object itself.

In a typical example, a doctor wishes to view inside the uterus of a pregnant woman in order to determine the size and shape of the fetus. This helps to: establish the stage of pregnancy, detect twins, show the baby's rate of development if intrauterine growth retardation is suspected, determine the baby's position in the uterus during late pregnancy, determine the position of the placenta if placenta previa is suspected (a condition where the placenta covers the fetus's head as it enter the cervix), or detect any fetal malformations.

Put briefly, ultrasound is the least-invasive machine technique available to obtain images or detect disease states of internal organs. The brief pulses of ultra-high-frequency sound waves act as a kind of submarine sonar, generating images from the sound echoes. Images are constructed from discontinuities in the speed of sound caused by the different media through which the sound waves pass, arising from variation in the density and the elasticity of the tissues that reflect and transmit these waves. From these variations, the sizes and shapes of organs are inferred based upon their acoustical properties. Typical applications include fetal-growth monitoring, tumor detection, and diagnosing abnormalities in the beating heart.

CHAPTER 18

Sights and Eyes

More than any other sense we possess, our sense of sight dominates our lives. We truly are creatures of the eye, and we even think in visual terms. We say, "I see," when we mean we understand something, and we hold to the old adage "Seeing is believing." The stage magician, knowing that we believe our eyes more than our other senses, fools us easily with optical illusions. But though our eyes can play tricks on us, vision is our most accurate sense.

The Physics of Light: A Boundary Between Classical and Quantum Physics

With normal vision, the eye is in continual movement. These movements are quite small and involuntary, persisting even when the eye is fixed on a target. As a result of this flickering motion, the image of the retina of the eye is also constantly in motion.

By using an ingenious technique, D. O. Hebb, Woodburn Heron, and Roy Pritchard, of McGill University in Canada, investigated what would happen if the results of that movement were canceled out. In other words, what would happen if somehow the image on the retina was brought to rest? To find out, they connected a tiny projector that projected an image onto a contact lens adhered to the cornea. Consequently, the lens would take part in the flickering motion along with the eye. In this manner, the image seen by the retina remained fixed, even though the eye and the lens flickered together; the lens moved with each eye movement.

The results confirm some interesting hypotheses. One is that image perception depends on experience. A man kept blindfolded from birth would not be able to make sense of things if the blindfold was suddenly removed. Another hypothesis suggests that no experience is necessary; the brain has separate, primitive patterns already established for visual perception. For example, the brain has a pattern that differentiates vertical lines from horizontal.

141

Seeing takes place as a result of "summing up" the separate primitive patterns and putting them together to make a whole picture.

The complete explanation of perception involves bringing together these two apparently contradictory views.

To get a better idea of the Canadian experiment, picture the subject lying down with the doctored contact lens in place in one eye, with the other eye taped shut. The subject is watching an image that appears on a tiny screen attached to the contact lens. The focus of the image has been so fixed that this image is in focus in the subject's retina. The image is like one produced by a tiny slide show—it remains fixed, unwavering, no matter how much the subject blinks or moves his eye.

The Canadian scientists found that in their stabilized image experiments, the image of a simple figure, such as a line, vanished rapidly, then reappeared. More complex figures also disappeared and reappeared, sometimes wholly, often in part. The length of time the image persisted depended on image complexity.

From the whole or gestalt hypothesis, the vanishing of the entire image can be explained. Once the brain "tired" of seeing the same thing over and over again, and not finding it threatening, it simply ignored it. The fact that the whole image vanished suggests that seeing can be explained in terms of the perception of a complete image. Conversely, the experiment also showed that at other times, the whole image did not vanish. Separate parts of the image would persist. For example, on seeing a picture of a cube, perhaps the vertical lines would disappear, and like the Cheshire Cat in *Alice in Wonderland*, the horizontal "smile" lines would remain. This supports the second hypothesis: Images are made from elements; the whole can be seen only as a sum of its parts.

These experiments, then, produced evidence that sustain both theoretical approaches to visual perception, and it appears that the two are complementary, similar to the way that wave and particle theories of quantum physics are complementary. According to the first theories of quantum mechanics, matter had to be composed either of waves added together or of separate particles. The quantum theory reconciled the two theories. However, that reconciliation removed quantum physics from the realm of immediate understading. Now both explanations exist side by side: Matter exists both as waves and as particles, depending on observation technique.

What determines whether one sees a wave or a particle in a quantum physics experiment? It depends on a choice made by the observer. If a wave is chosen, then the experiment must set up the apparatus so that a complete pattern of the waves can be seen. But in so doing, one cannot predict any single particle event. On the other hand, if a particle experiment is selected, the wave pattern is disrupted and one cannot see any wave influence at all. Of course, this is a gross simplification. In any one experiment, it is possible to find evidence of *both* wave and particle, never simultaneously, however.

The visual-perception experiment may be due to a similar duality; in fact, because it deals with light, it may indeed be the same thing. The whole pattern, when it is perceived, may be due to the mind's ability to perceive the wave aspects of light. This is not a simple one-to-one correspondence, however, between a single light wave and its perception. It may be due to the brain's holographic perception of many light waves, based on interference patterns, much as a film-emulsion hologram is constructed. The perception of the parts of the image may be caused by a particle perception apparatus in the brain. Again, no one-to-one correspondence is assumed. The mind sees separate parts because many separate photons—particles of light—are involved. Probably the truth is somewhere between these two extremes.

The parallel between the two experiments has another support. It is the quantum physics experimenter's choice which decides wave or particle. In the visual-perception experiment, the viewer determines whether he sees the whole pattern vanish or sees just separate lines.

Today, we are still somewhat in the dark about vision. Just how an image is perceived remains a mystery. Although we have identified the neural pathways excited in producing that image, the question remains, Where is the perceiver? Since visual perception apparently involves reconciling both a gestalt (or whole) image with its many parts, perhaps that reconciliation also follows the rules of quantum physics. In other words, the act of seeing must include the perceiver of the image. In quantum physics we have learned that the perceiver has a choice given to him by the wave-particle duality of matter. He can look at the wave, or gestalt, view of matter, or he can look at the "particle," or detail, view of matter. What he perceives depends on how he goes about looking.

In this chapter, I hope to explain a little about the mystery of vision—particularly the ability to see colors. It will turn out that color perception critically depends on a rather simple quantum process—one that is understood quite well. I also hope to show that color vision evolved over the millennia—and still is evolving and changing today.

These changes are more evidence of the body quantum in action.

Color Vision: A Quantum Process

The eye responds to stimuli within an extremely small range of vibrations of electromagnetic energy, known as the visual band, of the complete electromagnetic spectrum. These vibrations, like those created by sound waves, are characterized by two related concepts: wavelength, and frequency.

The complete electromagnetic spectrum exists over a wide range of wavelengths, roughly from .00005 nanometer, or smaller than the nucleus of an atom, to well over several miles. (A nanometer is a billionth of a meter.) The visual band can be described in terms of either frequency or wavelength, but it is best understood in terms of the sensation of visual electromagnetic energy that most of us experience. That sensation is, of course, color.

Every color has a wavelength and a frequency. Their relationship is particularly simple: frequency of light times the wavelength of light is constant (frequency × wavelength = constant). And that constant is simply the speed of light in the medium in which that light is traveling. This means that all colors of light travel with the same speed in any medium, including the vitreous humor filling our eyes.

The visual band of electromagnetic energy that we call light exists in the range of wavelengths ranging roughly from 390 nanometers (experienced as violet) to about 730 nanometers (red). Our eyes contain two kinds of light receptors. Because of their physical shapes, these special cells are called rods and cones. Rods are sensitive to colors within the yellow-green range corresponding to around 505 nanometers. Cones exist in three varieties: type A, or "blue," cones are most sensitive to 450 nanometers; type B, or "green," cones are sensitive to 525 nanometers; and type C, or "yellow," cones to light at 555 nanometers.

By the early 1920s, physicists had recognized a particularly paradoxical feature of light. It exhibits characteristics of both particles and waves. The wavelike characteristics can be described in terms of frequency and wavelength, but the particlelike characteristics cannot. Instead, physicists employ *position* and *momentum* to describe a particle—terms that refer both to the measurability of the location of a given particle and to the measure of its impact when that particle collides with another object.

Particles of light are called *photons*; they travel in straight lines in empty space at the speed of light, the same speed as light waves.

The connection between wave and particle characteristics was contained in two simple relations, $E = h\nu$ and $p = h/\lambda$, where E and p are the energy and momentum of the photon, h is a constant called Planck's constant, ν is the frequency of the light wave, and the λ, its wavelength. Thus, these relations describe the particlelike behavior on the left-hand sides of the equal signs and the wavelike characteristics of light on the right-hand sides. These relationships constitute the complementary nature of the physical world, a cornerstone of what is called the quantum theory or quantum physics.

Our eyes actually are quantum instruments capable of sensing both the wavelike and the particlelike behavior of light. In fact, they are designed in such a way that either description works without ambiguity.

When light travels from one place to another, it is often best to think of it as a wave characterized by wavelength and frequency. But when light strikes an object or is absorbed, it is best thought of as a particle having location and energy.

When light first comes into the eye, it passes through the lens. The ability of the eye to focus light upon the retina depends on the inherent curvature of the lens: the greater the curvature, the tighter the focus. The distance from the center of the lens to that point where light from a distant object comes into focus is called the focal length.

The physics of lenses is described best by the wave picture of light. Whenever light waves pass through a transparent medium, such as the lens of the eye or quartz glass, they undergo a wave phenomenon called *dispersion*. This means that the waves tend to move at different speeds through the medium, depending on their wavelengths. Thus, when white light (light containing many colors) passes through a lens, it breaks up into its different colors— much in the same way that water vapor creates a rainbow. Such dispersion causes the light to be poorly focused, and so the lens is not capable of focusing colors equally. Some colors come to a focus in front of the retina, others focus behind the retina. Using yellow light as the middle standard of focus, red light tends to focus in back of the retina, blue light in front.

With such a disparity in the lens-focusing ability, one would expect to see a blurring of colors. If you go to the movies often, you probably have been exposed to this blurring, especially during scenes aboard a submarine when the lights inside the sub are turned to safety red during an attack. The pervasive red light makes everything look blurry. Similarly, red clothes appear blurry on television. Yet only in extreme situations, when a color from one end of the spectrum predominates, will blurriness be a problem. In ordinary light situations, neither red nor blue light dominates and the eye focuses properly.

Once the light has passed through the lens and traversed the vitreous humor, it must hit the retina—and then the particle or photon concept of light is needed to describe the next stages in the seeing process.

The Quantum Sensitivity of the Human Photon Detector

How does a person become conscious of light? It appears that we are aware of light striking our retinas, even under very special circumstances, when as few as two photons strike. Somehow we are able to become aware of such a small number of distinct events. Recalling the quantum nature of the hearing thresholds and recognizing the evidence of light thresholds, we must conclude that both senses—and possibly all our senses—are quantum processes requiring conscious awareness. Somewhere lurking in the brain-retinal network there lies a conscious observer. Perhaps this observer turns up in the molecular framework of the nervous system itself, including the retina and optic nerve. The role of that observer is to change the wavelike nature of light into a particlelike event registering on the retina. This is the observer effect in quantum physics explained earlier. Without it the light would still persist in its wavelike guise, even after striking the retina!

Perhaps a brief explanation of the wave-into-particle effect is needed here. According to quantum physics, any particle of matter travels from one place to another as a wave spread out in space, like the wave emanating from a stone dropped into a still pond. The wave will continue to travel in this manner until some observer happens to interact with it. At that instant, the wave collapses to a single point, just at the location of the observer's instrument. Some physicists believe that the human mind interacts with the matter waves in our brain in a similar way: The mind observes and matter waves collapse. The human retina may represent the mind at work collapsing light waves down to single photons, which appear as spots on the retina.

In Table 10, the number of photons needed to cause consciousness of light in a photoreceptor rod cell are presented.

In order to be seen, light must behave in its photonlike guise. Photons strike sensitive cells called photoreceptors. As we see by looking at the table, the number of photons needed to create vision depends on wavelength. The rod is most sensitive around 500 nanometers, needing only about 10 photons to register. High sensi-

146

Table 10

Wavelength	Number of Photons	Sensitivity
400 (violet)	57	.0175
450 (blue-violet)	18	.0565
500 (blue)	10	.1000
550 (green-yellow)	20	.0505
600 (red)	192	.0052

tivity means that fewer photons are needed to cause that response. (These numbers are rounded averages obtained from several experiments.)

The fact that the rod receptor requires 192 red photons (at 600 nanometers) but only 10 blue photons (at 500 nanometers) to cause identical reactions is a direct measure of a quantum physical fact called probability. Rod cells are composed of a pigment called *rhodopsin*, which consists of 2 molecules: the *opsin*, and the *chromophore*. It is the chromophore that determines whether we see; specifically, it is a molecule capable of undergoing a quantum transformation that is sensitive to the photon energy.

If the energy of the photon is too high or too low, the molecule will tend to be insensitive to it, much like a resonant note struck on a piano. The note is produced when a hammer strikes a wire string tuned to produce a certain sound. If a very similar note is played, the string will resonate again—sounding along with the other; if a very different one is played, the string will not respond as well.

In quantum physics a similar effect is observed, only quantum physicists deal in probabilities of striking notes rather than resonant actualities. Photons with 500-nanometer wavelengths strike a resonance in chromophore, while photons with more or less of a wavelength are slightly off-resonance. Thus, the probability of a 400- or 600-nanometer photon causing the transformation is smaller than that for a photon with a wavelength of 500 nanometers. This means that it takes a greater number of 400- or 600-nanometer photons to induce the transformation.

Think of the photon-chromophore interactions as the toss of loaded coins: When a coin comes up heads, a transformation has occurred; tails means that the photon missed its mark. Suppose that the 400-nanometer photon-chromophore interactions (PCIs) are half-dollars, the 500-nanometer PCIs, quarters, and the 600-nanometer PCIs, dimes. The halves are loaded so that after 100

147

tosses, only 2 come up heads on the average; the quarters are loaded so that, again after 100 tosses, only 10 come up heads; the dimes come up heads only 1 out of 200 tosses. These numbers match the sensitivities found in the table.

Now fill up a number of bags with these coins, each containing n coins, either all halves, all quarters, or all dimes. The number n is varied from 1 to 200. The bags are opened and spilled out on the table. Clearly bags with only 1 coin will have little effect; those coins will most likely land tails. But as n increases, we find our first head appearing when about 10 quarters spill out of the quarter bag. Of course we also might see a head in the 6-quarter bag, or have to wait for the 13-quarter bag, but by the time we empty the 16-quarter bag on the table we are bound to see at least 1 head.

The first head in the bag of halves appears when we have spilled out around 50 halves, and the first dime head appears when about 200 dimes have been spilled out.

It appears to me that the sensitivity of the chromophore molecule is remarkably like our own individual sensitivities, and this similarity may be more than mere metaphor. Human sensitivity could be explained in terms of the probability of affecting certain sensing molecules in our bodies. Perhaps this sensitivity is what we mean and feel as human sensitivity.

In the world of the quantum, sensitivity means probability. In other words, how likely is it that the chromophore molecule will respond to *any* photon? In the case of 400-nanometer or 600-nanometer photons, the answer is, very unlikely if only one of those photons encounters the molecules; but if more than one manages to be in the molecule's neighborhood, that probability goes up.

Although it takes more 600-nanometer photons to fire off a receptor, once that receptor has fired it doesn't know what caused it to fire.

In fact, even if the photon wavelength is off, it can induce a molecular change anyway. Such is the world of quantum probabilities. It is even possible to make the chromophore "jump" without any light entering the eye at all! Any photons in the eye could cause the molecule to change. Even in complete darkness, the eye always has some photons present. By simply pressing on the eye, photoreceptors can be made to fire off just as if light had entered the eye. These firings are caused by photons created from the pressure of the aqueous humor on the retina.

As you can see, when a molecule responds to a photon it is simply incapable of determining what caused it to respond. To gain that information, more photons must be absorbed, and hence, sensitivity is connected to both photon wavelength and number of photons striking the retina. Consciousness then results in an equal appraisal of large numbers of "off" photons or small numbers of "on" photons, where *off* and *on* refer to how close the photon wavelength is to the sensitivity of the molecule. The mind, therefore, will perceive different events as the same, even though they were produced in different ways.

In this manner we may explain certain kinds of illusions such as hallucination. A different number of events other than visual photons coming from the outside world would tend to produce the same effect, much the way pressure on the eye produces the visual sensation of light entering the eye. Perhaps hallucinations result from neural signals coming from the brain and reaching the retina; if so, the person really "sees" the hallucination, just as a real image would be seen.

How Do We See Colors? A Quantum Physics Answer

Try to imagine what it would be like not to be able to see any color—to live in a black-and-white world instead. For many people, colorblindness is a reality. But what is color blindness? This term actually refers to a rather common condition of being unable to distinguish between certain colors. Literal color blindness, meaning the inability to distinguish any color—seeing everything as shades of gray—is extremely rare.

The mystery, then, is why is there color blindness? One answer is that color perception is an evolutionary change: Primitive humans and some animals could not distinguish one color from another. Perhaps in the early human adventure there was no need to see colors, and the ability to distinguish one object from another was enough. But then, as the environmental conditions changed, the ability to distinguish one animal from another, or perhaps one tribe from the next, became important for survival—and color vision filled the bill.

As yet, we have no firm answer to the evolutionary puzzle. What we *do* know is that color perception depends on quantum probabilities—the ability of photoreceptor cells to fire according to how sensitive they are to different wavelengths.

There are two types of photoreceptors in the retina: rods, and cones. There are around 120 million rods in each eye. They are distributed all through the retina, but tend to be maximally distributed at an angle of about 20 degrees off the central axis of the eye. Rods are called *scotopic*, which means used for night vision, and are indeed most sensitive when light intensities are low. Rods are useful at night when we are looking straight ahead and something approaches out of the dark just outside our line of sight. Light from the off-axis object strikes the rods that are also off axis on the retina.

Cones primarily are used in daylight, and they are responsible for our having color senses. Each eye contains around 6.5 million cones, distributed mostly within the 10-degree arc toward the nose of each eye, just off the center line. In the primitive eye, cones were probably most sensitive in the region around 550 nanometers (yellow-green). We can assume that this limited sensitivity was in some way adaptive. One might imagine the early atmosphere filled with dust and very hot. Picture our earliest-known ancestors foraging for food; by having their eyes sensitive to yellow-green, perhaps they could better distinguish plant life from the background—the first foods. But through human evolution three different cone sensitivities have appeared. Type A, or "blue," cones are most sensitive to 450 nm, type B, "green," cones to 525 nm and type C, "yellow," cones to 555 nm. The existence of these three different sensitivities offers a firm physiological basis for what is known as the trichromacy theory of human color vision—the theory that all colors can be created from only three colors.

We know today that associated with each wavelength of visible light—wavelengths in the range 400 nm to 700 nm—there is a single, or pure, color. However, our eyes are not able to sense differences in pure red at 630 nm from another red at 625 nm; a range of colors around 600 to 700 nm will appear the same to us. As the wavelength decreases to around 590 nm, we are able to see orange; at about 560 nm, we perceive yellow; dropping further, yellow appears at about 555 nm, green around 530 nm. Green persists until the wavelength drops below 500 nm, changing to blue at about 490 nm. Violet begins appearing at about 430 nm. The different shades of colors are due to mixtures of one color with another, not to the changes mentioned above.

A closer inspection of the above paragraph shows that we are relatively insensitive to wavelength changes at both ends of the visible spectrum, but quite sensitive to wavelength changes in the range from 500 to 600 nm.

According to the trichromacy theory, it is possible to reproduce all colors from just three primary colors; color television sets work according to this principle. For example, the eye is not able to distinguish pure orange at 590 nm from a mixture of red at 630 nm and yellow at 560 nm: The two taken together appear to us as orange. But why should two wavelengths, taken together, produce the same effect as one? Paradoxically, color vision is not at all dependent on the wave nature of light. If it were, we would be able to sense differences in the wavelengths of light and perceive those differences as colors. Thus, we would be able to tell the difference between a mixture of two colors and a pure color; for example, a blend of red (620 nm) and green (555 nm) would appear different from yellow (580 nm). The fact is that we cannot see any differences between the blend and the pure color itself.

Yet the mind resists the idea that color vision has nothing to do with the wave aspect of light. After all, each wave of light has a specific wavelength, and color is nothing more than the sensitivity of the eye to that wavelength. Red is anything from 600 nm to 650 nm, and yellow is around 580 nm, with blue and violet in the shorter range around 480 to 400 nm.

Indeed, if we plot energy versus wavelength for each color, we obtain a series of very simple curves.

As you see in Figure 10, each color corresponds to a particular frequency or energy associated with one wavelength. Thus, pure

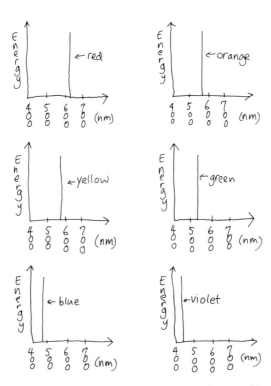

FIGURE 10. HOW COLORS HAVE WAVELENGTHS

151

red has all of its energy around 630 nm, while violet has all of its energy around 410 nm. As the wavelength decreases, the color we perceive changes.

White light, shown in Figure 11, is perceived as equal amounts of all colors. Experience has shown, however, that we see white when violet (480 nm) and yellow (580 nm) are blended together in equal amounts. White is also seen when red (600 nm) and blue (490 nm) are blended together in equal amounts.

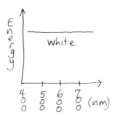

FIGURE 11. WHITE LIGHT ENERGY

How do we explain the confusion of white light with a blend of just two colors? The answer comes from quantum physics. As simple as the above curves are, a little thought will soon convince you that they have nothing to do with the perception of color as color! Remember that a photoreceptor, whether a rod or a cone, fires not according to the wavelength of the light (a wave property) but according to the probability of absorbing a photon. Once that photon is absorbed, all information concerning the energy or wavelength of that photon is completely lost. This is due to a principle of quantum physics called complementarity, which says that knowledge obtained by observing the particle nature of light must be at the expense of knowledge regarding its wave nature. Photons are perceived as particles; therefore, they cannot be seen as colors or wavelengths—it's a law of nature.

Thus, it would appear that color perception would be impossible not only when colors seem to cancel out into white light, but always. To explain the mysteries of color perception, we need a little more information about monochromatic vision—vision without color perception.

THE MONOCHROMATIC EYE

A hypothetical quantum sensitivity curve for a rod or cone, illustrating the response of a monochromatic or color-blind eye, is shown in Figure 12. The curve is an example of the well-known Gaussian, or bell-shaped, curve. Here, the curve indicates that the

152

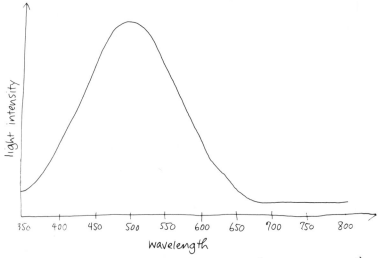

FIGURE 12. THE RESPONSE OF AN EYE'S CONE CELL (AN APPROXIMATION)

receptor is most sensitive around the color blue (500 nm). Thus, the probability of responding is greatest when a *blue* photon strikes a photoreceptor cell.

Remember that a photoreceptor cell fires regardless of the color of the photon that hits it. A 400-nm (violet) photon does just as well as a 630-nm (red) one, even though the probability of a red firing the cell is one-third that of a violet. In other words, any color photon has exactly the same effect on a photo cell: It fires, and that is that. Once fired, no memory is left of the color of that photon.

Consequently, if all cones were of equal sensitivity, we actually would be color-blind, and, indeed, all colors would appear as white light. As I pointed out earlier, I believe that early humans were probably color-blind, having only rods and monochromatic cones as photon receptors. If so, when we were cave dwellers, our cones would have been sensitive only to around 500 nm, responding as shown in Table 10 (page 147) and as illustrated in Figure 12. Suppose our ancestors looked up one day and saw a spaceship à la *2001* land, with flashing signal lights corresponding to two different wavelengths. Each second around 1,000 blue (500-nm) photons, and then, a little later, the same number of red (600-nm) photons, struck our cave person's eye. Could he tell the difference?

Yes, but the difference would be one of intensity, not color. Since the receptor responds to 10 blue photons the same way as it does to around 200 red ones, he would experience the 1,000 red

153

photons as simply a dimmer light than the 1,000 blues (1,000 reds produce the same effect as 50 blues). If the spaceship increased the number of red photons 20 times, all he could say was that the light was exactly the same as for the blue photons. But no color would be seen at all, only brightening and darkening white light, as in a black-and-white movie.

A DICHROMATIC EVOLUTIONARY EYE

But suppose that we evolve a few millennia, and now we have two kinds of receptors: one sensitive to red photons, and the other to blue. What could we tell then about color? Suppose that our two receptors responded according to Table 11 and as shown in Figure 13.

Table 11

1,000 Blue and 1,000 Red Photons

Wavelength	Receptor A	Receptor B
500 (blue)	10	100
600 (red)	192	35

If 1,000 red photons are received by the cave dweller's eye, 192 are absorbed by receptor A, but only 35 are absorbed by receptor B. But if 1,000 blue photons are seen, receptor A absorbs only 10, while receptor B absorbs 100. Thus, with two receptors, a difference would be perceived between the two wavelengths, a difference that would be perceived as color. In fact, all the colors could be sensed with just two kinds of receptors.

However, a kind of color blindness still persists—a confusion of white light with color mixtures. If a mixture of red and blue light shines in the primitive two-receptor eye, it causes both receptors to respond equally.

Each receptor will absorb about the same number of photons. This result, however, will be experienced as white light since there is no difference in the number of photons absorbed in each kind of receptor. Only when the mixture of colors is more red or more blue than the white light mixture will the cave dweller see any color. The movie is now sepia-toned, with some faded colors appearing.

Therefore, colors added together can produce the sensation of white, colorless light simply because they cause our receptors to respond equally. Can this equal response occur for just one wavelength? The answer is yes for the two-receptor eye. There is a

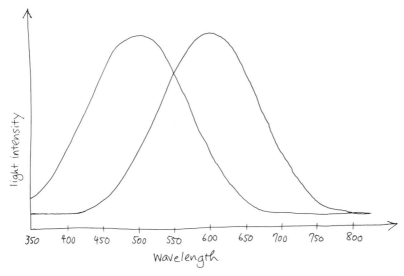

FIGURE 13. THE RESPONSE OF TWO DIFFERENT CONE CELLS IN THE EYE (AN APPROXIMATION)

particular wavelength where the same result as the mixture of colors will be experienced.

At a wavelength of 536 nm, which is somewhere around green, both receptors are equally sensitive, producing the same result as for a mixture of colors; see Table 12. The green wavelength where this takes place is shown in Figure 13; it is near the intersection where the two curves make an X.

Table 12

2,763 Green Photons		
Wavelength	Receptor A	Receptor B
536 (green)	210	210

This means that the two-receptor eye cannot tell the difference between green and a mixture of red and blue, and the eye perceives these as the same colorless white. However, the cave dweller who sees with only two kinds of cone receptors will not be totally color-blind, though he will tend to confuse green light with white light and suffer other color problems.

The two-receptor eye, or dichromatic system, is far from perfect. Our poor dichromate individual will not be able to distinguish any color differences in the 350- to 450-nanometer range

155

of wavelengths because there are no comparisons to be made between the two receptors. He will only notice the light getting brighter. Similarly, he will also fail to see much differences in colors for wavelengths greater than 650 nm, since above this level receptor A is asleep. Our dichromatic individual, as a result, is only able to see colors distinctly in the range from 450 to about 650 nm, and he easily gets confused when the wavelength is near 550 nm—he thinks it is white light.

THE TRICHROMATIC EYE OF MODERN HUMANS

Happily, evolution marches on, and now the modern eye contains three types of cone receptors, graphically illustrated in Figure 14. The graphs correspond closely to our own eyes but have been drawn more separately to better illustrate the point.

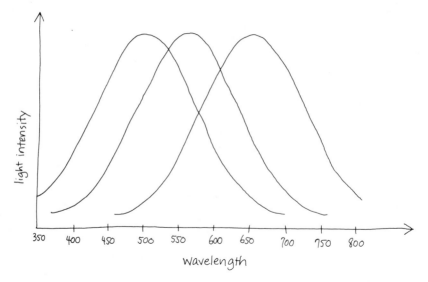

FIGURE 14. THE RESPONSE OF THREE DIFFERENT CELLS IN THE EYE (AN APPROXIMATION)

With three different receptors it is still possible to see white light, but now all three must be equally stimulated. However, no single wavelength can excite them all equally. Thus, no single wavelength will be confused with white, which is sensed when all three fire equally.

And what about color blindness? The most common form of color blindness today is the inability to distinguish between reds and greens in dim light. This is due to a defect in the sensitivity of

156

one or more of the three types of photoreceptors. Under brighter light conditions, these colors become distinguishable because there are more photons hitting these receptors. (Remember the example of the coins in the bags discussed earlier.)

This inability to distinguish colors appears to be a genetic throwback to our cave-dwelling days. Following the thesis that color distinguishability is an evolved characteristic, it would follow that some individuals would exist today with defunct type A, B, or C receptors, just as hairy noses and webbed feet appear from time to time to remind us of our early ancestry.

However, most thoroughly modern humans sense quite well in the extreme ranges of wavelength and can differentiate between shades of wavelengths from 350 up to 450 nm, where only receptors A and B fire. Our best color sensitivity exists when all three receptors are firing in the range from 480 through 680 nm. As far as we know, trichromatic color vision is the most sophisticated kind that exists in the universe.

But a little imagination tells us that if we were to develop a fourth cone, we would possess even greater powers of color differentiation. As evolution continues, it is quite likely that this will occur. Humans in the year 50,000 probably will see infrared and ultraviolet, and be able to distinguish between several blends of colors and single wavelengths. For example, the development of another cone, with a maximum sensitivity of around 400 nm, would enable humans to see many different colors in the blue to ultraviolet range with great sensitivity. There would appear as much color in this range of wavelengths as presently exists in the whole spectrum.

We are still evolving, still becoming more conscious, and more able to distinguish one thing from another. Indeed, evolution could be described as the ability to acquire distinguishability. That ability, certainly in color vision, is a quantum characteristic. Perhaps, as I pointed out in Chapter 14, this may be another illustration that evolution is a result of the body quantum.

PART FIVE

Breathing

I recently returned from a long visit to Brazil. I was invited to present a paper to a conference on mind-matter interaction, sponsored by IBM-do Brazil and UNICAMP at the University of Brazil located in Campinas, just 60 kilometers north of Sao Paulo. I found that the air in Brazil is quite pure these days, owing to the country's switch from gasoline to alcohol-burning engines. By 1990, all automobiles in Brazil will burn alcohol, which produces only carbon dioxide and water vapor as waste products of combustion.

But it was not only the clean air of Brazil that surprised me; vast portions of the country are covered by green jungle, producing, according to some estimates, more than 40 percent of the world's oxygen. If Brazil continues to cut back its "forest," as the Brazilians call their jungle, oxygen could become a rare gas. Perhaps Brazil could tax oxygen, or sell it to the world just as other countries sell oil and other natural resources.

Many countries today import oil, even food, to fuel their populations, but so far oxygen is still free. Yet we can all go without food for many days. A pause in the supply of oxygen of even just a few minutes can cause death.

Why do we need to breathe? The answer turns out to depend on a single subatomic particle, the electron. Following the laws of quantum physics, that particle is capable of both storing energy and releasing it through the chain of processes called the Krebs cycle. Without oxygen, that chain would be broken, and electrons would not be able to release the energy needed to support life.

What are the various organs involved in just moving that precious subatomic electron around? First, we see how the heart carries out its functions and learn how these functions are measured, using the principles of modern physics. We next look at how oxygen is circulated by the blood and how the laws of fluid dynamics provide insights into why our blood is the way it is. Finally, we examine all of the molecular processes involved in respiration to see what finally happens to that electron as it gives up the energy of life.

CHAPTER 19

The Heart and Its Measurement

According to the latest surveys, cardiovascular disease is the number-one killer, and within that categorization lies the great fear: heart attack. To allay that fear, physics now offers a way to diagnose abnormal heart function before irreparable damage is done. In the not too distant future, we may even be able to obtain EKGs by just a few simple connections of electrodes coupled to hand-held microcomputers. It is toward this goal, as well as for the purposes of general education about the heart, that I have written this chapter.

The EKG: How It Works and What It Measures

At the heart of electrical measurements of the body is the electrical measurement of the heart itself. The measurement is part of a field of electric heart measurement known as *electrocardiography*, which literally means making an electrical map or graph of the heart. The key insight into how such a measurement is carried out lies in understanding the physics of heart muscles.

THE PHYSICS OF HEART MUSCLES

The heart is a muscle, and before a muscle cell contracts, a wave of electrical depolarization sweeps across the cell. This wave is quite similar to the wave of electrical depolarization that sweeps along a nerve cell's axon. When such a wave occurs, electric fields are generated outside the cell; these fields produce differences in voltage, usually in the range of thousandths of a volt, which are measurable as *electromyograms*—the electrical graphs of the *myocardium*, the muscle of the heart. These voltage differences exist over a wide area of the whole body, and they are relatively easy to measure.

It might appear amazing to some readers that the heart hidden inside the chest produces an electric field outside itself, and to

161

grasp how this is possible, we need to look again at the physics of electric fields.

The Electrophysics of Heart Cells

Myocardial cells and nerve cells are very similar. With both, a membrane separates an extracellular fluid medium from an intra-cellular fluid medium. The imbalance in the concentrations of the ions of sodium, potassium, chlorine, and large negatively charged protein molecules in these media is about the same across the membrane (see Chapter 15). The fluids on both sides of the cell membrane are also electrically neutral. One difference between nerve and muscle cells is that nerve cells do not actually touch each other; they communicate across synapses, while myocardial cells send electrical signals to one another at actual points of contact.

The heart is really a double pump containing four chambers (see Figure 15). Venous blood enters the two top chambers of the heart. They are called the *atria* (the singular is *atrium*), and they fill up through one-way valves that allow blood to enter but not to leave. This keeps the atria closed until pumping action begins. Then blood simultaneously moves in both atria through the re-spective valve gates into the two lower chambers, called the *ventricles*.

Blood enters the heart through two main veins called the *superior* and *inferior vena cava* on the right side of the heart, the *pulmonary vein* on the left. The vena cava carry deoxygenated blood from all parts of the body, while the pulmonary vein carries fully oxygenated blood coming from the lungs.

Blood leaves the heart from the ventricles through two main arteries. Deoxygenated blood leaves the right ventricle through the pulmonary artery and returns to the lungs for oxygen. Oxygenated blood leaves the left ventricle through the aorta to return to the rest of the body.

The heartbeat consists of a double-phased action. With each beat, both oxygenated and deoxygenated blood pass through the heart. In the first phase of relaxation, called the *diastole*, the atria and the ventricles fill with blood. As the ventricles relax, valves in the aorta and the pulmonary artery close with a "dup" sound and blood pours into the two atria from the vena cava and the pulmo-nary vein. Next the mitral and tricuspid valves, between the atria and the ventricles, open, allowing blood to enter the ventricles. The heart then stops filling with blood.

In the second phase of pressure, the heart undergoes a contrac-

FIGURE 15.

A SCHEMATIC VIEW OF THE HEART AND THE FLOW OF BLOOD

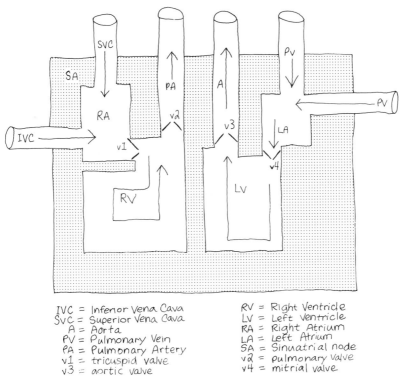

IVC = Inferior Vena Cava
SVC = Superior Vena Cava
A = Aorta
PV = Pulmonary Vein
PA = Pulmonary Artery
v1 = tricuspid valve
v3 = aortic valve

RV = Right Ventricle
LV = Left Ventricle
RA = Right Atrium
LA = Left Atrium
SA = Sinuatrial node
v2 = pulmonary Valve
v4 = mitrial valve

tion called the *systole*. The blood now empties into the aorta and the pulmonary artery, and a "lub" sound is made by the closing of the mitral (bicuspid) valve on the left and the tricuspid valve on the right. This prevents a backflow of blood into the atria. The whole action lasts around .8 second, with the diastole and systole each lasting about half that time.

Interestingly, if the heart is removed from an animal and bathed in a suitable nutrient solution, it will continue to beat completely spontaneously. With each beat, a wave of depolarization sweeps over the heart and contracts it. The beat-controlling factor is contained in the right atrium, referred to as the *sinuatrial* (SA) *node*—also known as the pacemaker. The SA node fires at regular intervals about 72 times per minute (though exercise and strong emotion increase the rate, while meditation and rest decrease it). The electrical signal from the SA node initiates the depolarization of both atria, causing them to physically contract, squeezing inward and exerting pressure on the blood contained within them. The blood is then pumped into both ventricles, and the atria are repolarized. The electrical signal next passes into the *atrioventricular* (AV) node, which initiates the depolarization of both ventricles. This causes the ventricular cavities to squeeze in, exerting pressure on the ventricular blood and forcing it out into the pulmonary artery (from the right ventricle) and the aorta (from the left ventricle)—the main artery out of which oxygenated blood is distributed to all the cells of the body. The ventricles then repolarize and the process repeats.

Normally, the AV node spontaneously depolarizes at a rate of 50 beats per minute, but since normal heart rate is higher due to the SA node pattern, this depolarization is rarely observed. Highly trained athletes, however, particularly long-distance runners, often possess extremely low SA beat patterns in their resting hearts because of superb physical conditioning. They consequently experience spontaneous AV depolarizations. Thus, for example, if an athlete has a resting heart rate of 40 beats per minute, he would experience 10 extra AV depolarizations, measurable as nodal escape beats. These beats would appear on an EKG as random heartbeats superimposed on the regular SA pattern. These beats even feel different because they are not in step with the regular pattern of beats; they are "extra," offbeat pulses. Clearly, the AV beat is not affected by exercise. It depends only on the basic electrical and chemical properties of the cells, and appears to be a kind of safety valve, ensuring that the heart beats regardless of the SA firing pattern. Some athletes have even become alarmed at the "extra" beats of their hearts, but these physiological beats are no cause for concern.

By now we see that there is a rhythm and progression to the heart's depolarization—literally, a wave of depolarization crosses the heart.

Electric Heart Waves

If we were to look closely at the heart wall while it beats, we would see the wave of depolarization sweeping down the heart wall. Originally, the cell is completely polarized. For each plus charge on the outside cell wall there is an equal minus charge inside the wall. The presence of the boundary and the distribution of the charges causes the electric field to vanish on both sides of the membrane.

However, as soon as a depolarization wave starts, an electric field actually is created along the direction of the wave's progression. This may not appear logical at first, since as a result of the wave, charges of electricity have vanished—seemingly removing sources for that field; however, the dynamic quality of the wave motion actually creates a measurable field at the body surface.

Measuring the Heart Fields

The whole movement of the heart field can be read by a voltmeter, an instrument that detects electrical voltage differences. By attaching electrodes to the skin at strategic points, the field can be measured as a function of time. The reason that such a measurement is possible can be understood by examining the field produced by the heart.

This field actually extends in space well beyond the heart. Much like the magnetic field of a bar magnet, the electric field appears and extends quite a distance, depending on the field strength. The usual way a physicist portrays the field is in terms of lines of electrical force as shown in Figure 16.

FIGURE 16. MEASURING THE ELECTRIC HEART (EKG)

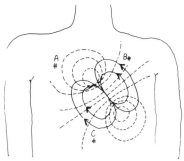

The electrodes usually are placed at points near or at locations A, B, and C. Most commonly, they are placed at the left arm (LA), the right arm (RA), and the left leg (LL). The electric field produced by the heart wave is shown by solid lines in the figure. The field actually extends in three dimensions, and can be visualized from the figure by imagining rotating the butterfly-shaped loops around the line running diagonally through the heart.

The broken lines in the figure show what physicists call equipotential lines. Along these lines the voltage has the same value. As time marches on, the field and its equipotentials change, causing the voltages measured at the electrode points to change correspondingly.

Now, the voltage between any two points actually measures the strength of the field. In order to determine its strength absolutely, measurements must be made along different lengths; that is why more than two points are used. In principle, if the voltage differences are measured three times along three mutually perpendicular directions (the famous three dimensions of space), the field strength will be known. Sure enough, during a standard EKG, 12 different potential voltage measurements are made, using 9 electrodes.

Since the field is produced by a repeating wave of depolarization across the heart, the field traces out a repeating pattern across the space of the body. That spatial pattern shows three distinct regions called P, QRS, and T, shown in Figure 17.

FIGURE 17.
THE TIP OF THE HEART'S ELECTRIC DIPOLE

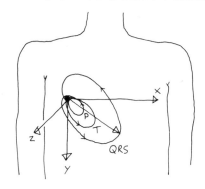

It is this pattern that reveals whether or not a heart is beating wildly, mildly, or ineffectually. That pattern, traced out by the electric field, reveals the health of the heart. The field line, shown by an arrow in the above figure, must follow a particular route, and, like a bus on a schedule, it must arrive at locations P, QRS, T, and back again to P exactly on time. If it does not, something is wrong with the heart.

The electrodes attached to points A, B, and C (see Figure 16) on the body pick up the "electric bus" on its journey, reproducing it as a graph called an EKG.

One from the Heart Field

A typical EKG is shown schematically in Figure 18.

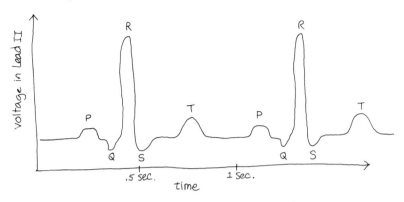

FIGURE 18. WHAT AN EKG LOOKS LIKE

This shows roughly the potential voltage difference between the right arm and the left leg. Actual EKGs do not appear much different from the schematic unless there is an abnormality in the heart beat.

The P wave (strial depolarization) appears as a tiny hump, signaling the start of the heart beat. The Q signal next appears, indicating the beginning of the ventricular depolarization. As this phase continues, the field increases in strength, reaching a maximum at R where the field actually points in from the right arm down to the left leg. At S, the field is actually pointing in the opposite direction, from C to A. As the heart repolarizes, we pick up the smaller T signal.

Hearts: A Future

We have witnessed in the last several years three remarkable heart surgeries: the famous Barney Clark artificial heart, the baboon heart surgery on Baby Fae, and the artificial heart surgery on William Schroeder. Although both Barney Clark and Baby Fae died, their sacrifice gave us much-needed knowledge. We know more about the limits of artificial hearts—what they can do, and what we shouldn't expect. Present thinking holds that artificial hearts should be used only as emergency pumps until a donor human heart can be found.

Mr. Schroeder has as of this writing suffered a setback caused by a stroke—a blood clot in the brain. This clot has been directly attributable to the problem of interfacing mechanical tubes connecting the artificial internal mechanical heart to the main pulmonary artery, the aorta, the inferior and superior vena cavas, and the pulmonary vein. These are the five major inlets and outlets to the heart.

Achieving a correct interface poses a problem in physics having to do with the onset of turbulence and the dynamics of blood flow near and around obstacles. The problem is magnified because blood is a living fluid, and living cells do not necessarily follow the same patterns of movement as do non-living inert particles.

While artificial-heart research is still very much in its primacy, the hope remains that the ultimate model will feature a miniaturized drive system permanently installed in the chest, with a portable battery pack worn around the waist.

A CYBORG TRADE-IN

Suppose we extrapolate to a cyborg future, where human beings routinely trade in their hearts and other organs as needed. What would a typical futuristic mechanical heart look like? It would probably be made of soft plastic, resembling the actual heart in size and bulk. It too would have four chambers, and its walls would be capable of sustaining electrical charges. In this manner it would produce EKG patterns nearly identical to those of the normal heart. Installed within its boundaries would be a tiny electric pulse maker, replacing the SA node. An externally manipulable AV-node control would be installed so that the spontaneous heartbeat could be lowered to match such unusual environmental conditions as long sleep or the prolonged absence of air.

How would this heart be used? To extend life, certainly, but also to extend endurance to new levels. Remember, the heart's primary purpose is the circulation of oxygen. Cells can go without food much longer than they can go without oxygen. For the cyborg of the future, his artificial heart and oxygen supply would enable him to explore places, perhaps as dangerous as outer space or under the sea, without the use of pressurized suits. Even more useful, this artificial heart could be monitored from a distance: A computer-controlled central "brain" could keep track by radio signal of all artificial-heart recipients. Such signals would be amplified from the natural signals produced by the P, QRS, T "electric bus" tripping through the chest cavity of each cyborg's body.

The amazing thing about all this is that this future is within our grasp now; the technology for radio-controlled artificial hearts exists today.

CHAPTER 20

Living in Pressure Suits

The term pressure has actually become a metaphor, another name for the stresses of everyday life—work demands, family crises, health problems, and other problems. But what is pressure, from the point of view of physics?

What Is Pressure?

It certainly is the most common form of force we experience. Each night, for example, the TV weathercaster tells us the atmospheric pressure; we apply air pressure to our tires; if we feel run down, we go to the doctor's office and have our blood pressure checked. Our bodies could not operate without pressure! We use pressure to breathe; pressure exists inside our skull, in our blood vessels, our heart, eyes, digestive system, throughout the skeletal system, and, on the microlevel, across the vital membranes separating our cells.

Pressure is defined as force acting across an area. A typical man's shoe has a heel with about 6 square inches of area. When a 160-pound man takes a walk, he applies all 160 pounds on each heel for each step. This means he exerts a pressure of 160/6 pounds per square inch, or about 27 pounds per square inch. A 130-pound woman wearing "sexy" high heels with .25 square inch of heel area, aside from the awkward posture she assumes, applies a pressure of more than 500 pounds per square inch with each step. No wonder that some linoleum floors are marked and dented by women's shoes; ouch, I don't want my toe underneath that!

Because we live at the bottom of a "sea of air," we feel the pressure of that air. Air has weight: A column of air just one square inch in cross-sectional area extending upward 5 miles (where air itself peters out) weighs around 14.7 pounds.

Pressure is measured in units of force per area. This could be newtons per meter squared, pounds per square inch, or dynes per square centimeter. However, in medicine and weather, pressure is usually measured with a column of mercury (Hg is the atomic symbol for mercury.) Mercury, being much denser than air (around 13,300 dynes per cubic centimeter compared with 1.3 dynes per cc), doesn't need to be as high as 5 miles in order to produce a pressure of 14.7 pounds per square inch (or 1 atmosphere) at the base of its column. To yield 1 atmosphere (atm) of pressure, a column of mercury need rise only 75 centimeters, or 750 millimeters.

Thus, if the atmospheric pressure were to decrease, that column of mercury would fall a few millimeters; an increase would cause it to rise a few millimeters. When the weathercaster speaks of a falling or a rising barometer (pressure indicator), it literally means a falling or a rising column of mercury.

Since we exist at the bottom of this sea of air, all pressures we measure actually are relative to 1 atmosphere of pressure. The familiar tire pressure of 30 pounds per square inch is actually 30 + 14.7, or 44.7 pounds per square inch absolute pressure. The reason we consider relative or *gauge* pressure is that inside our bodies we find pressures lower than 1 atm. Such pressures often are referred to as negative pressures (meaning less than 1 atm). When we breathe in air, the pressure in our lungs must be less than 1 atm or else air would not enter. Sucking up soda pop through a straw also produces negative pressure.

Blood Pressure: What It Measures About You

With each beat, the heart acts as a pump increasing pressure in the arteries. A typical or normal blood pressure, usually measured by a cuff surrounding the upper arm, is 120 mm Hg peak (or systolic) pressure. In other words, when the heart reaches its peak pressure, that pressure would drive a column of mercury 120 mm higher than atmospheric pressure alone. Just after reaching peak pressure, the pressure diminishes to its "background," or diastolic, pressure; typically, this pressure is around 80 mm Hg. In a blood pressure reading, both the systolic and diastolic pressures are determined and written as 120/80, indicating the range of pressure experienced in the body.

According to many physicians, high blood pressure is a killer. The systolic pressure can rise as high as 140 mm Hg and still fall within the range of normalcy, but a diastolic pressure beyond 90 mm Hg is in the danger zone. As your heart pumps blood through the arteries, the blood flow exerts pressure on the arterial walls. With an increase of diastolic pressure, the arterial walls must stretch, thereby increasing tension. Similarly, the heart itself must increase its muscular tension to keep up the pressure, which, in turn, forces blood to flow faster through the system.

According to the physics of fluid flow in a tube, such as blood flow in an artery, the fluid flows faster where the tube is narrow, slower, where wide. Correspondingly, the pressure is higher in the wider part of the tube, lower in the narrower. If an obstruction appears in an artery, it causes the blood to flow faster in order to get the same amount of blood around that obstruction. The heart must therefore produce more pressure to maintain that faster flow. Indeed, blockages or narrowing of the arteries is often the reason that the blood pressure increases in the first place.

To grasp this concept, we need to look at a well-known formula from classical physics called Bernoulli's principle.

BERNOULLI'S PRINCIPLE: FLYING, BASEBALLS, AND BLOOD FLOW

We all know that a baseball curves when thrown by a talented pitcher. We also know that airplanes fly and that shower curtains grasp our legs annoyingly as we stand under the stream of water. During a high-speed automobile drive, a sudden roll-down of the window will often suck maps and other articles out the window. And as I described above, whenever an arterial cross section is narrowed, the blood flow increases in speed as the blood pressure in the narrowed artery decreases.

The baseball's curve and the sudden whoosh of maps sucked out of a car's window also result from the laws of physics applied to fluids. In these examples, the fluid is the air that surrounds the baseball as it flies and that rushes through the car window as the car cruises down the highway. All fluids such as the air sustain pressure differences and corresponding flow velocities, which attempt to equalize those pressure differences. The familiar north wind that freezes us in the winter and the balmy south wind that warms us in the summer are examples of fluids flowing under pressure differences.

Each of these phenomena, including blood flow, is an example of the old physics principle named after its discoverer, Daniel Bernoulli (1700–1782). Briefly, it states that the pressure in a fluid decreases with increased velocity of the fluid. When a fluid flowing steadily in a wide pipe encounters a narrow constriction, it must increase its speed in order that no fluid backs up, disrupting the steady flow. If the fluid meets this constriction in the form of a narrow exit mouth, the fluid compensates and flows out faster. The familiar garden-hose nozzle squirts in accordance with Bernoulli's principle.

That the pressure decreases in the narrowed constriction is not so obvious, however; the reason has to do with work that is being done on the flow by pressure in the tube. At the wide end of the tube, pressure is what moves the fluid in the direction of the flow. Fluids flowing in tubes also have energy. In fact, Bernoulli's principle arises from a well-known law of physics—the conservation of energy. Pressure can be regarded as a form of potential energy, while the flow of fluid constitutes a form of kinetic energy. Taken together, kinetic energy plus potential energy must remain constant in each part of the tube. That is why the flow is faster in the narrow part, slower in the wider. The potential energy (pressure) in the wider part is greater, while its kinetic energy (flow velocity) is less. Just the reverse is true for the narrow part.

A good way to think about blood pressure is in terms of potential energy. High pressure is more potent than low, able to release more kinetic energy, like a high-pressure streamer or locomotive; too high, and something's got to give.

THE HEART PUMP

When the blood leaves the heart through the aorta and the pulmonary artery, it travels at about one foot per second. As the blood branches into smaller and smaller arteries (called *arterioles*), that speed decreases to nearly a standstill. The reason, again, is Bernoulli's principle. The total cross-sectional area offered to the blood as it branches has been increased from around 3 cm^2 to nearly 600 cm^2. The flow speed, consequently, has dropped to around 1 mm per second—a snail's pace.

We can understand the slowing of the blood as it branches out from the heart as a simple consequence of Bernoulli's law. Because the blood flows from a large-diameter artery into many smaller ones, having a total cross-sectional area much larger than the main artery, the blood velocity in each of the smaller arteries must decrease. This is, however, based on the assumption that the blood flow is steady—which it is not. Consequently, the blood pressure throughout the system of arteries, veins, arterioles, and capillaries actually varies considerably. Typical pressures in an artery are around 90 mm Hg; in a small capillary, this figure drops to about 30 mm Hg; and in a small vein, the pressure is about 15 mm Hg.

These figures at first may appear to contradict Bernoulli's law—that whenever blood flows from a large-diameter artery into a smaller one, it must flow faster, and that whenever it flows from a smaller one into a larger, it correspondingly must flow slower. Blood does behave this way. The difference here is that the blood flows from one main artery, with a given diameter, into many

172

smaller arteries that have a total cross-sectional area greater than the original. The net effect is that it flows from an artery with a smaller cross-sectional area into a complex of arteries having a greater total cross-sectional area.

Don't be fooled by the fact that the secondary arteries have smaller cross-sectional areas than the primary one. The blood must flow slower as it branches out from the heart because there is just so much blood to go around; if it flowed faster, it would create a vacuum in the primary.

The key to grasping all of this is common sense: There is only a certain amount of blood, and it must flow, more or less, steadily through the system. When the total cross-sectional area increases, the blood slows down in order to allow the faster blood behind it time to fill in the space. When the total cross-sectional area decreases, the blood must flow faster in order to avoid a traffic jam of blood coming in from the larger diameter behind.

It is in the small capillaries that oxygen and carbon dioxide exchange. The low speed of blood flow is what makes such an exchange possible, because the gas-diffusion rates increase when the fluid that the gas enters is slow-moving.

Suppose that, each second, your body requires a certain volume of blood to flow in order to bring oxygen from the lungs to the cells and to take carbon dioxide away from the cells, returning it to the lungs to be dispelled. The flow rate will depend on the pressure difference produced by the heart. In general, the flow rate is a linear function of the pressure difference in the direction of flow.

The flow rate also will depend on the thickness of the blood. Honey flows slower from a bottle than champagne. A measure of the ease that a liquid flows is called the viscosity. Blood is three times as viscous as water, thus giving rise to the adage "Blood is thicker than water." An increase in red-blood-cell count will increase the viscosity of the blood, as will a drop in body temperature. Cold blood is thicker or more viscous than hot.

POISEUILLE'S LAW AND ATHEROSCLEROSIS

The French physician Jean Marie Poiseuille studied blood flow in the 19th century. He began his studies by looking at the flow of water in tubes of different sizes. He is also credited with the *poise*, the unit of viscosity named after him that equals 1 dyne-second per square centimeter. Poiseuille found by experimentation just how the flow rate varied with changes in viscosity, tube length, different pressures, and tube radii. His findings indicated that:

- Doubling the pressure doubles the flow;
- Doubling the viscosity halves the flow;
- Doubling the length halves the flow; and
- Doubling the radius increases the flow by 16 times.

In determining these facts, he held all other variables equal to their original values; that is, when he doubled the pressure he kept the viscosity, length, and radius of the tube the same to see what effect it would have on the rate of flow. Poiseuille's results also can be described according to how variations affect the pressure:

- Doubling the flow doubles the pressure;
- Doubling the viscosity doubles the pressure;
- Doubling the length doubles the pressure; and
- Doubling the radius decreases the pressure by 16 times.

The surprising factor is the large changes in flow and pressure produced by varying the tube radius. Since doubling the radius decreases the pressure by a factor of 16, halving the radius would cause an increase in pressure by the same factor. Thus, Poiseuille discovered that in order to maintain a certain flow through the tiny arterioles, a much larger pressure change would be needed. Without such change, the small radii of the arterioles would cause so great a drop in flow speed that an actual stoppage would result. To compensate for such a stoppage, the heart would have to work overtime to increase the pressure; that would result in overtaxation.

To grasp the significance of this data, we need one more piece of the physics of fluid flow: the difference between smooth (or laminar) flow and turbulent flow.

The Onset of Bloody Chaos: Turbulence in the Flow

Go wading in a smooth-flowing stream on a quiet summer day high in the mountains, and you will understand the physics of laminar flow. Near the shore, the water is hardly moving at all; as you wade in deeper, you'll feel the flow of the stream more with each step. In the center of the stream, the flow is quickest.

Similarly, laminar blood flow is most rapid in the center of the blood vessel; the layer of blood nearest the vessel wall moves slowly. This change in the distribution of blood flow affects the distribution of red cells, causing more red cells to be in the central stream than toward the sides. Since arterioles branch off from the sides of an artery, fewer red cells will be skimmed off.

174

Now let us consider an artery with a continually narrowing cross-sectional area: The tube radius decreases as the blood flows along its length. As the tube narrows, the flow speed increases (according to Bernoulli's principle). At a certain point along the tube, the speed of the flow will reach a *critical velocity* when smooth flow suddenly changes into turbulent flow.

The onset of turbulence depends on the radius of the tube, the viscosity of the fluid, and the fluid density. It also depends on a certain number called the Reynolds' number. In 1883, Osborne Reynolds discovered that:

- Doubling the viscosity doubled the critical velocity;
- Doubling the density halved the critical velocity;
- Doubling the radius halved the critical velocity; and
- Obstructing the flow decreases the critical velocity.

In the aorta, which has a radius of around 1 centimeter, the critical velocity is .4 meter per second (m/sec). The velocity of blood in the aorta ranges from 0 to .5 m/sec, and, thus, during the systole or high-pressure phase of blood pumping, the flow is actually turbulent. Indeed, it is the turbulence of the flow that we hear when we hear blood squirting through the arm arteries during a blood pressure measurement.

There is a critical difference in blood pressure between laminar and turbulent flow rates. A laminar flow is more sensitive to a change in pressure than is a turbulent flow. This means that for the same change in pressure, there will be a greater change in laminar flow rate than in turbulent flow rate. When the pressure in a normal artery increases, the flow rate correspondingly increases until the flow rate reaches a critical speed. The pressure at which this speed is reached is called the "critical pressure." As the pressure increases even further beyond the critical, the flow rate increases only slightly.

On the other hand, if an artery is obstructed for any reason, the pressure needed to produce a given flow is correspondingly increased. Because of the obstruction, the onset of turbulence takes place at a lower flow speed, according to Reynolds' discovery. But because of the increased pressure in the obstructed artery, the critical pressure is actually higher. The net effect: An obstructed artery has a lower critical velocity and a higher critical pressure than a normal artery. Since the obstructed artery must pass the same amount of blood as the free-running artery in order to supply the body's needs, a much greater pressure will be produced in the obstructed artery than in the normal one. To produce the higher pressure, the heart needs to work harder. Unfortunately, all arteries will be faced with the increase in pressure produced by the obstruction of just one artery, and blood pressure will rise.

The Dangers of High Blood Pressure, According to Physics

As we get older, the laws of classical physics still prevail in the hydrodynamics of our blood flow. Arteries tend to accumulate fatty tissues, called *atheroma*, at certain locations, particularly at artery branches, where, as we have seen, laminar blood flow is slow. If you want a dramatic demonstration of atheroma formation, consider the ordinary household electric fan. Even after the fan has been blowing all day, dust accumulates on the fan's blades because the air flow against the fan's blades is zero just at the blades' surfaces. Similarly, in our arteries, wherever the circulation is slowest, the fat is thickest.

As atheromas grow, they obstruct the arterial flow, resulting in the onset of the disease which is the main killer of us all: atherosclerosis. As we age, the walls of the arteries also become less resilient to pressure changes. This is called arteriosclerosis (hardening of the arteries). Thus, as increased blood pressure occurs incrementally over the course of our lives, the walls of the arteries become less elastic, less able to change. Instead of an elastic response to a pressure increase, eventually the vessel can simply "crack" or break open. When this occurs in the brain, it is called a stroke. Also, atheromas are more likely to form where strokes have occurred already. (Strokes are also caused by a blood clot that tears loose from the wall of a vein and migrates to the brain.)

One method of treatment is lowering the blood pressure through the use of drugs or biofeedback, thus removing the risk of damage caused by pressure blowout. High blood pressure not only causes damage in the arteries and veins, it can also damage organs. Organs such as the kidneys, the brain, and the eyes can work only within certain tolerances of blood pressure. Further damage can occur in other organs if high pressure causes a breakdown in a specific organ. For example, high pressure can damage the kidney, which, in turn, because it is not operating properly, tends to increase the blood pressure even further, eventually causing brain or optic-nerve damage.

HEART WORK

The work done by the heart follows the fundamental gas law of physics, relating pressure and volume change:

work = pressure × volume change

The average pressure in the heart is roughly halfway between the systolic and diastolic pressures, around 100 mm Hg, or 1.4×10^5 dynes/cm². (A dyne is an extremely small amount of force. One drop of water, about one ml, weighs nearly 1,000 dynes.) Each second the heart pumps around 80 ml of blood (about 2.72 ounces); in less than 12 seconds, a whole quart of blood has been pumped, and in one minute, 1.32 gallons of blood have passed through the heart. In that short time period a single red cell has managed to make a complete round trip from the heart all the way around the body and back again to heart. In one day, the heart pumps 2,709 gallons of blood.

Since the volume change is 80 ml per second, the heart performs $80 \times 1.4 \times 10^5$ ergs of work each second, or about 1.1 watt-seconds of energy. It is rather amazing that such a seemingly small amount of energy accomplishes such a large amount of work! However, the heart beating inside our chests is like a 1-watt light bulb that never turns off. If you think about it a little more, you will realize that the electrical power we use in our homes requires a rather large expenditure of energy. For example, even assuming that electric generators are 100 percent efficient (meaning that all of the energy used to run the generator is converted into electrical power), each watt-second of power would require a mechanical pump pushing 1.32 gallons of water each minute if that pump produced only a tiny pressure of 1.4×10^5 dynes/cm². One dyne, remember, is really a very small force.

A 100-watt light bulb would need 100 heart-sized mechanical pumps. All together, 132 gallons of blood would pass through them each minute. Of course, if each pump were able to produce a greater pressure, fewer gallons of blood would be needed.

However, most electrical generators are not 100 percent efficient. And, correspondingly, most devices, such as the heart, that convert electrical energy into mechanical energy are also not 100 percent efficient. The lower the efficiency, the more electrical energy required. Because the heart is only about 10 percent efficient, the average heart's power consumption is nearer 10 watt-seconds.

The work done by the heart is roughly the pressure or tension in the heart walls times the volume change per second times the amount of time the heart beats. The rise of blood pressure increases the heart muscle tension. A speeding up of the heart rate also increases the work load. With increased work the heart itself needs more oxygen and thus needs its own blood supply.

It is here that the heart is likely to run into trouble: heart disease, and its effect, heart attack. A heart attack is caused by a blockage of one or more of the arteries supplying oxygen to the heart muscle. The major arteries supplying that precious breath of life are called the *coronary arteries*. These, in turn, branch out over the surface of the heart.

Coronary artery disease is common in the Western world, where it accounts for about 30 percent of all deaths. When your heart beats faster in response to physical or psychological stress, it requires increased oxygen; severely narrowed or blocked arteries cannot cope with this need.

Air Is Not Free; The Work of the Lungs

Breathing in air requires energy. We breathe in about 6 liters (a little more than 6 quarts) of air every minute. Men on the average take about 12 breaths per minute, while women take about 20, and infants 60. The air we breathe contains about 80 percent nitrogen, 20 percent oxygen; the air we expire contains the same amount of nitrogen (we don't get our nitrogen from the air), 16 percent oxygen, and 4 percent carbon dioxide. We also expire roughly .5 kilogram (around 1 pound) of water vapor each day. Thus, in just breathing out you lose 1 pound of water a day.

Just as blood leaving the heart finds itself entering smaller and smaller blood vessels, where it slows down and drops pressure allowing diffusion to occur, air, after passing down the trachea (the throat's windpipe) enters two bronchi, one in each lung. From there the bronchi split into *bronchioles*, which split further into the small sacs called *alveoli*. Each bronchus divides and redivides about 15 times, resulting in alveoli of around .2 millimeter in length. These alveoli look like tiny interconnected bubbles only about twice as thick as a sheet of paper; the bubble walls are only .4 millionths of a meter thick. Each bubble expands and contracts with air, just like tiny balloons. In them, oxygen and carbon dioxide exchange—the real reason for breathing.

Breathing in and out can be likened to tides—and, in fact, the normal .5 liter of air men breathe each second (.3 for women) is called the *tidal volume at rest*. With effort it is possible to extend the volume of air taken in or expired. The lungs possess an inspiratory volume of about 2 liters, with an expiratory reserve of around 1.2 liters. This reserve keeps the lung from collapsing—a serious condition that occurs only in damaged lungs. The normal lung is never empty of air, not even after the full, extended exhalations of yoga breathing. The full range from peak inspiratory volume to peak expiratory volume, including the tidal volume of air, amounts to 4.4 liters of air.

Typically, normal breathing uses up about 2 percent of the total energy consumed by the body. Using simple concepts from classical physics, it is possible to make a mechanical model of the lungs (see Figure 19). The whole respiratory system of lungs, chest wall, and diaphragm can be imagined to consist of simple physical elements: a weight or mass, m; a spring with a force constant, k; and a dashpot with a resistance, r.

FIGURE 19. A MODEL OF THE LUNG-CHEST DIAPHRAGM SYSTEM

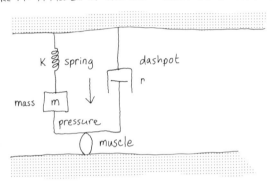

Think of the resistance, r, as the resistance to the motion of lung tissue and the flow of gas into the lung. The spring, k, represents the elastic response of the lung-chest-diaphragm system together. The mass, m, represents the inertia of all the moving parts of the breathing system. When the muscle contracts, pressure is exerted downward; when it elongates, the spring pulls the whole system back. This corresponds to the downward and upward movement of the diaphragm.

Mass, m, responds to forces by offering resistance to acceleration, called *inertia*. The spring responds to forces by offering resistance, k, to stretching. The dashpot offers resistance, r, to motion.

When you breathe, all of these resistances come into play. The inertia, m, and the spring force constant, k, trade off energy as the lungs expand and contract. When the lungs are contracted as the air is exhaled, the spring is compressed slightly, causing energy to be stored in that compression. The lungs momentarily are not moving, but the tension in the spring is high and you are "dying for a breath." As the spring reaches its midway length, air is filling your lungs and there is no energy in the spring. But the lungs are now moving and will keep moving as more air comes into them. All the energy of the system is found in the motion of the lungs (kinetic energy) and there is no acceleration. When the lungs are fully expanded, they have stopped moving and the spring is stretched to its maximum.

During all this process, the dashpot is resisting with a force dependent on the velocity of the lung movement. This resistance produces heat and dissipates energy. Thus, the faster we breathe, the greater the motion in the lungs, and, because of the increased resistance, the more energy is used up.

Closely observe your breathing and you will notice that it is easier to breathe in than to breathe out. This is because during inspiration, forces in the airways tend to open them further, while just the opposite occurs when we breath out. Patients suffering from obstructive airway diseases, such as asthma or emphysema, suffer even more when they attempt to breath out quickly. They usually find relief by taking short breaths, which keep the lungs relatively full of air and the air passageways opened as large as possible.

During normal breathing, no work is done during expiration. In the model the spring snaps back, air is expelled, and the dashpot burns up some energy. While you are strenuously exercising, however, your muscles must expire the air and work is done. In fact, during heavy exertion such as long-distance running, breathing may use up 25 percent of your body's total energy consumption.

Rapid shallow breathing and slow deep breathing are also wasteful of energy. Returning to the model, the spring tends to respond to motion at a certain frequency only; this is called the resonance frequency, and it is easily observed by watching a vibrating spring. Rapid or slow breathing is either too fast or too slow in comparison to the resonance frequency. Low breathing rates cause work to be done against the elastic forces of the spring itself; fast breathing brings the dashpot into the picture, causing resistive forces to produce heat. All in all, the model corresponds closely to real breathing systems.

Contrary to popular thinking, aerobic exercise such as running will strengthen the muscles but will do little to improve lung capacity. Any increased stamina is due to the strengthening of the muscles.

Indeed, we live in pressure suits with a remarkable system of physical laws governing the regulation of the pressures throughout our bodies. As we have seen, some very basic laws of physics govern pressure, whether it is the air pressure in our lungs or the blood pressure in our veins and arteries. Without the "pressures" of life, no life is possible.

CHAPTER 21

The Quantum Physics of Blood

A quick glance through this chapter may cause you to wonder why it is entitled "The Quantum Physics of Blood" and not the chemistry of blood. Chemistry deals with such things as the structure of molecules and how molecules combine with each other. In Part Three, we saw that quantum physics is the physics underlying all chemistry; thus, blood chemistry is really blood quantum physics.

In this chapter we will look at blood in its primary capacity as the transport or circulation system for oxygen. However, because blood types do exist, I also have included some brief explanations about them and why problems arise when a donor with one type of blood attempts to give to a recipient with another type. You could think of the blood system as a railway train. If there are cars missing, you would like to know how to fix the train up, and when you lose blood, like a train with too few cars, the only way to fix it up is to add more. But better be sure you add the right kind: Some cars won't couple to others, and some blood types cannot be given to others.

The Circulation System

No matter how simple or small an animal is, it still needs to breathe to ensure that it supplies itself with oxygen and eliminates carbon dioxide and other gaseous wastes. This means that the gases must cross a barrier—the skin of the animal—in a process called diffusion. The greater the surface area of skin exposed to the gases, the greater the volume of gases actually diffused.

A typical one-celled animal has a ratio of nearly 14,000 square feet of skin surface area per pound, while a human being has only around .1 square foot per pound. With this much surface exposed to the outside world, it is no wonder that, in the case of single-celled animals, ordinary gas diffusion accomplishes the task of gas exchange.

These simple, single-celled animals are able to exchange oxygen and carbon dioxide—that is, breathe—through their skins. Being so tiny, they accomplish this through the advantage of having a large surface-to-body-weight ratio. As you may recall from an earlier chapter, the surface-to-body-weight ratio of an animal decreases as the body volume increases. Thus, large animals lose less heat to the outside air than small ones. That is why, for example, polar bears are larger than their brown bear cousins. The polar bear lives in a colder climate than the brown bear, and needs the smaller surface-to-weight ratio in order to minimize the heat lost through its skin to the frigid outside air. But having less skin exposed per pound leads to the problem of oxygen supply and carbon dioxide elimination for each cell. To accomplish this, large animals need a circulation system; hence, they need blood.

Being as complex as we are, and not possessing the enormous area-to-weight magnitudes of simple cells, we too need an internal circulation system to bring needed oxygen to every cell in our bodies. And evolution has provided us with an internal waterway system which cleverly transports gases from the outside world to each and every cell in the body, and back again. Blood is the river of human life's transport system. It is capable of taking in oxygen simply by being exposed to air; it also is capable of holding on to it through the magic of quantum physics called blood chemistry. It can also pick up carbon dioxide from the cells and deliver it back to the lungs.

This process of transporting gases in the body is initiated by the simple act of breathing. The expansion and contraction of your lungs actually sets in motion a complex series of quantum-physical interactions. Blood is pumped to the lungs from the heart under the relatively low pressure of about 20 mm Hg, or at about 15 percent of the circulatory pressure of the body overall. Your lungs act as gas exchangers, with carbon dioxide expelled from the blood solution, while oxygen dissolves in it, combining with molecules in the blood. To accomplish this exchange effectively, the lungs must contain many nooks and crannies, offering to those gases a large total surface area on which to assimilate and expel. Indeed, the surface area of the lungs, if stretched out, would cover half of a tennis court!

Iron Richness—How Oxygen Combines with Hemoglobin

Blood contains the oxygen-transport protein called hemoglobin. The main component of red blood cells, hemoglobin chemically combines with oxygen in the lungs and transports it to the body cells. In the cells, oxygen combines with glucose—the main food supply—to produce energy; this process is called *aerobic glycolysis* (see Chapter 22), or, simply, oxidation. Hemoglobin is referred to as a respiratory pigment because it evolved from the same molecules that give us skin color. If you remember, most smaller animals breathe through their skins, which, in turn, suggests that we may have evolved from a smaller species of animal. Perhaps our ancestors' outer skins contained respiratory pigment that became enfolded and internalized, producing dissolved hemoglobin—skin molecules in solution. Maybe that is how blood evolved in the first place. In addition, for oxygen to reach the hemoglobin in the lungs, it must diffuse through a layer of water. We all breathe through water, in other words, which suggests that the species from which we evolved also lived underwater.

Physiologist L. J. Henderson once described hemoglobin as the second most interesting substance in the world; first prize went to chlorophyll, the green pigment molecule in plants. Remarkably enough, the two molecules are nearly identical. Perhaps both plants and animals have an ancestor in common, and a common function associated with color. The only difference is that in the hemoglobin molecule, iron occupies center stage, while in chlorophyll, it is magnesium. And it is to iron that we now turn our attention.

A normal human being contains about four grams of iron, 70 percent of which is distributed in hemoglobin. Around 5 percent of the body's iron is found in myoglobin (the material making up muscle), 5 percent is found in special respiratory enzymes, and 20 percent in the manufacturing and storage areas of the bone marrow, liver, and spleen. A tiny percentage, only four milligrams, is found floating freely in the bloodstream, not bound to hemoglobin. This tiny amount is a clue to a process occurring over and over again in the bone marrow—the manufacture of new hemoglobin molecules from the dregs of the old. The body actually salvages iron from destroyed red blood cells and circulates it in the blood as free iron salts, after which it is taken up by the bone marrow to make new blood cells.

But the most important use for iron is in the hemoglobin molecule. There, an iron atom takes a central location in a latticelike structure of molecules called the heme. The heme appears as shown in Figure 20 (page 184); also shown is the chlorophyll molecule (Figure 21).

At the center of the scaffolding-lattice system lies a single iron

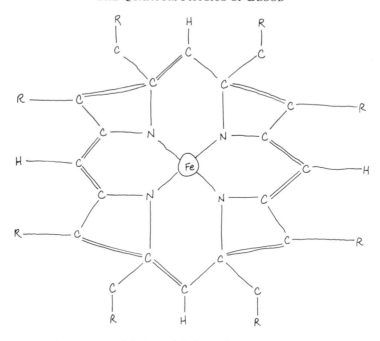

FIGURE 20. THE SKELETON STRUCTURE OF HEMOGLOBIN

atom (Fe) in a ferrous state, which means that it is doubly ionized—electrically charged with two protons, Fe^{2+}. Surrounding it are four nitrogen atoms. This scaffolding all lies in a single plane.

Also attached to the Fe ion, but not shown because they lie much farther away, are four large spiral protein chains making up the globin of the hemoglobin molecule. The main function of the spiral chains is to protect the oxygen molecule from the forces of the water molecules surrounding it. These chains interlock to form an ellipsoid, with the dimensions of 64 × 55 × 50 angstroms (10 billion angstroms = 1 meter) and a molecular weight of about 67,000 (so many times the weight of one hydrogen atom). Each chain is linked to the heme, forming a cradle in which the oxygen is drawn. The oxygen molecule is nicely tucked into the cradle of the hemoglobin structure, and the four long spiral chains are made of polypeptides, or amino acid groups (see Figure 22).

It is important that water be kept away from the ferrous ion; otherwise, oxygen in the water would act on its electrons, leaving the iron in a triply ionized state and preventing it from binding to

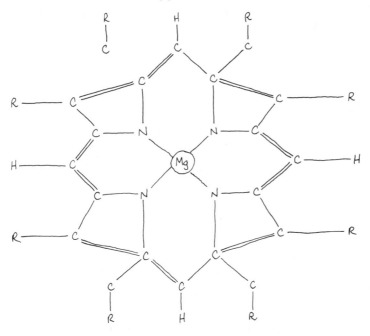

FIGURE 21. THE SKELETON STRUCTURE OF CHLOROPHYLL

oxygen in the blood. That is the reason for the oxygen cradling provided by the spiral globin chains.

When you breathe, your hemoglobin molecules breathe also. The oxygen you take in attaches to the iron, and the whole molecule changes its structure to accommodate it.

Today, the screening of human blood samples in many parts of the world has led to the discovery of nearly 100 different mutant hemoglobins. These mutations all appear to be harmful, producing such symptoms as inclusion body anemia, a condition where hemoglobin precipitates out of solution forming bodies in the red cells and shortening their lives. Another symptom is cyanosis, a blue color of skin due to the presence in the capillaries of deoxyhemoglobin (hemoglobin without oxygen). The reason for this condition appears to be that the ferrous ion has an affinity too low for oxygen—it cannot hold it powerfully enough. A third symptom occurs if the affinity for oxygen is too high, so that the hemoglobin does not release its oxygen to the cells. This is called polycythamia, and it causes the marrow to overproduce red cells in compensation. (Perutz and Lehman.)

FIGURE 22. A STEREOSCOPIC VIEW OF THE HEMOGLOBIN MOLECULE

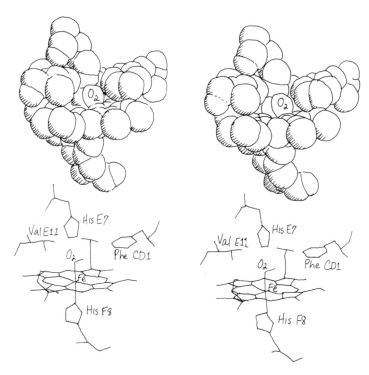

Focus your eyes so that your left eye views the left picture, and your right views the right picture, so that a three-dimensional image appears in the middle. (Adapted from S. Phillips, 1980 Journal of Molecular Biology.)

Blood Groups and Their Quantum Differences

Many of us will need to receive or to give blood sometime in our lives. However, there are certain restrictions governing who can give blood to whom, restrictions based on subtle differences in the quantum physics of blood cells. I want to explore these differences here. Perhaps a comment about blood transfusions, however, is necessary before I begin. Blood transfusions are quite common, vital to human survival, yet there are a few frightening aspects to be considered.

Today, many of us are concerned about the possibility of contracting the disease **AIDS** through blood transfusion. It appears that the **AIDS** virus is indeed transmitted through the intimate

186

contact of body fluids, and blood appears to be a good conduit. Yet every day many thousands of patients require blood transfusions, and the Red Cross still conducts drives for human blood. With precaution, I am sure that donating blood and receiving it are safe. Of course, the problem of receiving blood that contains mutant hemoglobin molecules—a kind of quantum mechanical disease—would not normally be detected. Detecting molecular changes requires sophisticated equipment yielding X rays of such high resolution that the molecular structure can be seen. However, such mutations are rare and should not be of concern to you if you need blood.

Why do such mutations occur? Why are there any differences in molecules of human blood at all? The answer appears to be evident from quantum physics—the body quantum in action. A mutation is nature's way of changing life, of experimenting with new forms of life that may evolve into something better suited to a new environment. To accomplish this task, according to my quantum physical view, molecules will "consciously" alter themselves, perhaps in response to their environment. Remote quantum physical probabilities suddenly can become realities. According to quantum physics, nothing is absolutely certain, and new possibilities will always arise eventually. Again, the conscious aspect is contained in the observer effect—the ability to change a possibility into actuality simply by observing it.

Through such actions in the distant past, different blood groups probably originated. Although red cells in different people look much alike under the microscope, they are, in fact, dissimilar. They can be divided into four distinct groups: A, B, AB, and O. The two most common groups in western Europe and the United States are A and O: about 43 percent of the population is A, another 43 percent, O. Group B accounts for around 10 percent, AB, less than 5 percent. The difference in blood groups depends on molecules carried on the surface of the red blood cells called *antigens* and on the presence or absence in the plasma of other molecules called *antibodies*.

The different blood groups, therefore, are classified according to the presence/absence of antigens and antibodies. An antigen is a molecule capable of creating an antibody; an antibody, in turn, is a molecule capable of targeted attacks on other specific molecules. In fact, when you receive an inoculation, you are getting an injection of antigens, which create antibodies that provide continuing protection against such bacterial diseases as polio. This process goes something like this:

Antigen — (creates) → antibody — (destroys) → antigen

And it repeats over and over again. The body recognizes the process and is able to produce more antibodies in defense of further antigen attack. The body actually is introduced to a new disease through injection of the antigen. In response, the body goes on to produce more antibodies than the number of antigens first introduced, which gives the body a head start against further bacterial infection.

Antibodies act by causing antigens to stick together, rendering them less effective in exchanging materials from the bloodstream in which they float. This decrease in effective exchange is due directly to the decrease in the surface-area-to-weight ratio of the cell clump: the larger the clump of antigens, the less their effect on the body.

There are two antigens, A and B, and two antibodies, $A^\dagger$ and $B^\dagger$. In various combinations, these antigen and antibody factors produce the four blood classifications or groups.

If blood plasma (the clear liquid making up 55 percent of blood volume) or whole blood from one group is introduced into the veins of a person of another blood group, a dangerous reaction, called *agglutination*, will occur. The way this works can be seen by looking at the different blood groups described next.

Again, the four blood groups are labeled A, B, AB, and O. Group A blood contains A antigens and $B^\dagger$ antibodies that attack B antigens; group B, B antigens and $A^\dagger$ antibodies, attacking A antigens; group AB has both A and B antigens, but no antibodies; while group O blood has no antigens but has both $A^\dagger$ and $B^\dagger$ antibodies. The antigens, A and B, are molecules attached to individual blood cells. The antibodies, $A^\dagger$ and $B^\dagger$, are floating freely in the blood plasma.

To understand who can give blood and who can receive it requires looking at what takes place when blood from one group is introduced to another group. The important question is: What happens to the donor red cells? If the recipient's blood contains antibodies that will attack the donor's antigens, the blood cannot be given. If the donor's blood is antigen-free, no danger exists, even though the donor blood may contain antibodies. Why? First of all, the absence of donor antigens means that the recipient's antibodies will not attack the donor red cells. Second, even though donor antibodies are present, they will do little damage because the relative volumes of donor to recipient blood are so different. The recipient's blood contains far more red cells than the donor's antibodies can handle.

According to the foregoing, can you determine who can give blood to whom? Take an AB recipient, for example. Since she has AB blood, she has no antibodies in her blood plasma. This allows her to receive red blood from anyone! On the other hand, her blood does contain both A and B antigens, and she therefore cannot give her blood to any group that has A† or B† antibodies. This means that she can give her blood only to another AB recipient.

Now consider an O blood type. His blood contains no antigens but has both A† and B† antibodies. Poor O cannot receive blood from any group other than an O donor. But generous O can give his blood to anyone without fear of agglutination occurring.

The Rh Factor

In addition to the above blood groups, there are several other classifications—Rhesus, NNS, P, Kell Lewis, Dufy, and Lutheran, to name just a few. But by and large these classifications are minor factors and most individuals do not possess antigens or antibodies that fall into them with the exception of the Rhesus, or Rh factor.

The important Rhesus blood group system was discovered in 1939, when antibodies were produced by injecting red cells from a rhesus monkey into a rabbit. The antibodies that developed in the rabbit's blood were found afterward to agglutinate not only with rhesus monkey cells but also with a high proportion of human beings as well. Most important, the blood serum from women who had given birth to babies suffering from what is called hemolytic disease of the newborn was found to cause agglutination in the same cells, although the ABO grouping was compatible.

This meant that human blood may also contain antigens and antibodies of the Rh factor. The presence or absence of the Rh antigen makes every individual either Rh positive or Rh negative. About 15 percent of the population is Rh negative; 85 percent contain the Rh antigen molecule on their red cells.

The major problem occurs when a pregnant mother is Rh negative and her baby is Rh positive. During the last few weeks of pregnancy, Rh positive cells from the baby escape into the mother's bloodstream through the placenta. While causing no problems in the first pregnancy, in subsequent pregnancies, particularly the third and fourth, agglutination may occur in the baby's blood. This is because the mother will have had time between pregnancies to build up Rh antibodies. Since they are in her blood plasma, these antibodies will invade the subsequent baby's bloodstream, leaking back across the placenta and causing destruction of the baby's red blood cells.

About 17 out of 100 women possess Rh negative blood. If the husband is also Rh negative, then all the offspring will be Rh negative and no problem develops; if the husband is Rh positive, occurring statistically in 14 out of the 17 cases, the baby stands a 50 percent chance of being Rh positive. Thus, for every 100 pregnancies, there are about 7 births in which an Rh negative mother produces a baby who is Rh positive.

If an Rh negative father is matched with a mother who is Rh positive, which happens in 12 out of 100 cases, there is no problem because the mother does not have Rh antibodies present. If the baby is born with Rh positive blood, no problem occurs; if the baby is Rh negative, the invasion of Rh negative blood cells into the mother's Rh positive cells prevents antibodies from being generated. With Rh negative blood, the baby normally will not have Rh antibodies because the mother's red cells do not cross the placenta to reach the baby's bloodstream. Thus, no antibodies will be generated.

CHAPTER 22

The Quantum Physics of Respiration, or Why Do I Breathe?

Over countless centuries of evolution, life has developed remarkable ways to take in and use materials from the environment and convert them to energy. We are all familiar with our need for oxygen: We must breathe, and oxygen is needed for the production of energy from food.

Anaerobics and Aerobics: The Breath for Life

As I mentioned in Chapter 5, all processes that require oxygen in order to proceed are called aerobic (airlike), and those not needing oxygen are called anaerobic (not airlike), or fermentation processes. The glass of wine you enjoy with dinner was made by anaerobic processes; in fact, the presence of oxygen needs to be avoided at all costs in the wine- and beer-making industries. Now you know why your wine is tightly corked.

When life first began, oxygen was *not* plentiful in the atmosphere. In order to live, cells had to take in food and make energy anaerobically. Remnants of our early beginnings still exist today, mostly in the form of microorganisms that live in soil, deep water, or marine mud. The pathogen (disease-causing) organism *clostridium welchii*, which causes gangrene in wounds, is an anaerobic life form.

Some of our cells, such as skeletal muscle cells, can adapt to oxygen deprivation. They are called *facultative* cells. When oxygen is absent from their environment, they can extract energy from glucose by anaerobic, or fermentation, mechanisms. When oxygen is present, these same cells show a marked preference to burn glucose instead—to oxidize it.

Oxygen first appeared, undoubtedly, when plants learned how

to photosynthesize—to build up glucose from sunlight, water, and carbon dioxide, with oxygen as a waste product. (Remember that when a cell utilizes glucose without oxygen present, the process is called anaerobic *glycolysis*.) Our muscle cells are capable of anaerobic glycolysis, and, in fact, this is the reason why muscles become tired. The runner who attempts the 100-yard dash or 200-meter run must put his muscle cells through anaerobic glycolysis simply because he cannot supply enough oxygen to his muscles in the period of time needed to run the race at top speed. The tiredness is caused by the production of lactic acid—the natural by-product of anaerobic exercise.

The chemistry of lactic-acid buildup is simple: Glucose converts to lactic acid plus 47 calories of energy. This translates into about 119 calories per pound.

With the presence of oxygen, however, the reaction changes: Glucose plus oxygen converts to carbon dioxide plus water, plus 639 calories.

As you can see, the burning, or oxidation, of glucose provides many more calories (around 1,614 per pound), compared with glucose fermentation, with lactic acid as the product. Consequently, with an aerobic process lactic acid is not produced, and more than 13 times as much energy is created.

CANCER CELLS HATE TO BREATHE

It is known that cancer cells can live both aerobically and anaerobically. They seem, however, to prefer to live anaerobically, burning up large amounts of glucose through fermentation rather than taking in oxygen, even when it is available. One of the key symptoms in cancer is loss of body weight, and this phenomenon can be understood in terms of the greater need for cancer cells to consume glucose anaerobically. Each cancer cell produces only one-fourteenth the amount of energy per gram of glucose compared with a normal aerobic cell. Thus, glucose depletion quickly occurs and the body must go after its fat deposits to supply its energy needs. Cancer is a sure-fire means for weight loss.

Three Steps in the Process of Life

The complete oxidation of glucose is the most efficient way to convert food into energy. The process involves several intermediate steps. Aerobic glycolysis is the first in the three-step process of life's energy production; these steps are:

- Aerobic glycolysis;
- Krebs cycle; and
- Electron respiration.

During the first step, glucose is converted into pyruvate, which is needed by the second step, pyruvate representing the end product of aerobic glycolysis. Next, pyruvate is converted to citrate, which is used in the Krebs cycle as fuel.

The Krebs cycle functions like a wheel rolling through mud; with each turn of the wheel, energy must be added to overcome the friction of turning the wheel through the mud. With each turn, particles of mud are also thrown off, taking more energy from the turning wheel.

The form of energy needed to "turn the cycle" is citrate. This molecule is converted via a series of steps into oxaloacetate, which, in turn, becomes energized to produce citrate all over again. The "mud energy" thrown off by the Krebs cycle is the change in form of a particular molecule called nicotinamide adenine dinucleotide, or NAD for short.

NAD provides the basis for the third step, electron respiration. Electrons which have been added to NAD during the Krebs cycle are taken away during this step and given to oxygen. In a sense, oxygen "breathes in" electrons, so this process is called electron respiration.

Here the real energy exchange occurs because the electrons given to NAD during the Krebs cycle have high energy that they release during the respiratory step.

When oxygen breathes in electrons, another, very important molecule, called adenosine diphosphate (ADP), becomes energized. When it becomes energized it takes on an additional phosphate molecule, becoming adenosine triphosphate (ATP). As I described in Chapter 5, ATP is utilized in nearly all of the body's cellular processes as an energy carrier. It gives up this energy by dropping off its added phosphate cargo, converting back to ADP. Thus the energy cycle of the body can be based on the simple cycle:

$$ADP \rightarrow ATP \rightarrow ADP, \text{ and so on.}$$

THE RED OX AND THE ELECTRON

An important ingredient in the breathing cycle is the molecule nicotinamide adenine dinucleotide. To grasp the role of NAD, we need to look at what is called the "red ox" of chemical reactions. When a particular molecule gives up an electron, it is said to undergo oxidation, and in so doing it passes from the reduced, or "red," form to the oxidized, or "ox," form. Thus, Xred becomes Xox plus e−. Usually the electron is carried away from the Xox by another atom or molecule.

The oxidation process normally takes energy, while the reduction process, whereby an electron is gained, liberates it. A simple example of this process can be seen in the oxidation of a hydrogen atom: When hydrogen is oxidized, it changes into one proton plus one electron; when the proton is reduced, it changes back into hydrogen once again. Oxidizing hydrogen requires energy; reducing protons gives it off.

During the Krebs cycle, a molecule of NAD in its reduced form, NADred, is produced. Energy is created when protons interact with NADred, a complex reaction producing two hydrogen atoms and NAD in its oxidized form—NADox. This molecule, it turns out, carries a positive electric charge, like a proton, and there is also something else missing: a hydrogen atom.

The net result is that when NADred is converted into NADox, energy actually is liberated instead of absorbed. That is why NADred is useful, and it provides the starting point for the third step in the breathing process.

Although this electron-respiratory step is not understood completely, it appears certain that electrons from NADred are transported from one important enzyme to another. The enzymes are called cytochromes, and they are actually red in color when viewed under a microscope, resembling hemoglobin. As the electrons flow from enzyme to enzyme, the energy they carry is released to upgrade ADP molecules into ATP.

The three cycles of breathing life are shown schematically in Figure 23. During the glycolysis step, glucose passes through several stages on the way to becoming pyruvate. Between two of these intermediate stages, F6P and FDP, a controlling mechanism, known as the ATP block, is exhibited; it also is seen during the Krebs cycle. This block inhibits these steps just at certain stages, as shown in the figure. Thus, should too much ATP be present, then F6P is inhibited from turning into FDP. Similarly, too much ATP will slow isocitrate change into ketoglutarate.

In this manner, energy provides its own control: Too much ATP and the system slows down. This means your metabolism slows down too, because the body gets the signal that you have enough energy stored in ATP molecules. This mechanism explains why successful dieting is so difficult. When you reduce food intake, your system starts burning body fat; this produces an excess of ATP molecules, and your metabolism rate slows down. In this

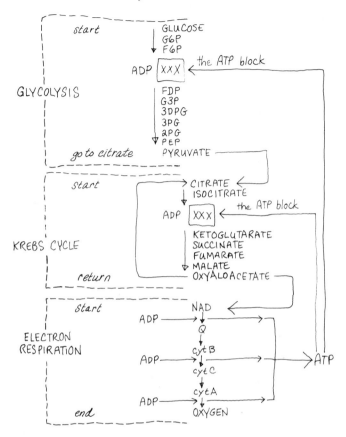

FIGURE 23. THREE CYCLES OF LIFE

manner, the fat cells tend to monitor their size: They don't want to reduce too much; they protect themselves from further reduction. By increasing your physical output through exercise, you can increase your metabolism, forcing the fat cells to let go of even more fat.

Why Do I Breathe?

For two reasons: first, to provide a better pathway for glycolysis. When oxygen is present, glucose produces pyruvate needed for the next step, the Krebs cycle, instead of lactate or lactic acid, which leads to muscle fatigue. Second, oxygen is the landing point for electrons as they make their way down the final step of electron respiration.

The sum total of all of this is that you breathe to change 36 molecules of ADP into 36 molecules of ATP. The energy-rich ATPs then are carried to every single cell in the body, providing the very energy of life itself.

PART SIX

Minding

So far, most of the processes we have looked at have literally been mindless—they are carried on with little or no mindful activity. Our understanding of how the mind enters into the processes of the body is still imperfect, but we do know that the mind plays a very active role. We know that we can alter our breathing rate simply by willing it. We can move the body from one place to another. We know that thoughts affect the body, producing anything from extreme pleasure to serious pain, even disease. In fact, there is some evidence indicating that stress brought on mentally may be the cause of the premature death of brain cells.

One thing is certain: The mind plays no role without a nervous system. Somehow that system conveys information from place to place in the body, and changes it into the intelligence process of cognition. Cognition requires that we know something about the information carried by the system; that is, to cognize is to recognize. To recognize something you must already have information with which to compare it, and somehow this information, or memory, is contained in the brain, muscle, and nervous systems. But just how, no one actually knows.

In the following chapters, I offer speculation about this mysterious mind-body interaction, and introduce yet another role for the body quantum. It is through this role that the mind alters the probabilities of events, turning them into the realities of everyday life.

The final chapters of this book address the concepts of minding, healing, and transforming; and they are controversial. I speculate about and offer some serious philosophical insights into the mechanisms of the mind. However, I caution the reader that much of what I say here is speculation about mind-matter interaction and how quantum physics indicates just what the connection is.

For example, consider the question of what constitutes a thought. (I include both conscious and unconscious mind processes in the thought category.) In my view, thought starts out as unconscious association; through several associations, a consciousness eventually emerges. Consciousness, in other words, starts out in the unconscious and eventually surfaces as a conscious thought. By that point it is too late to turn back, for whatever physical processes ensue with consciousness (and I believe that the most important process is the observer effect) will have occurred by the time we become aware of them.

My model of the mind consists of a supermind that acts as an integrative device: It integrates—adds up—the content of many smaller minds, each of which acts, through the observer effect, by choosing an actuality from a vast sea of possibilities. It is in this manner, for example, that molecules enter into certain quantum states. Then the supermind integrates molecular contents and votes. It is this vote that finally is registered as the conscious thought that results in a conscious action. I hope to clarify this concept as we continue.

CHAPTER 23

Communications: You Got Your Nerve

As organisms increase in size and complexity, from the one-celled amoeba to the multicelled human being, networks of internal communication become progressively more problematic. It is a wonder, in fact, that we don't fall all over ourselves, like the hapless centipede of joke fame who can't mind its many legs.

But perhaps the humble bumblebee is a better illustration of the problems involved in communication. It is sometimes wryly noted that a bumblebee cannot fly because its wings plainly are too small for its body. But the bumblebee has no use for aerodynamic theory and flies anyway. The reason it can fly is not a problem of flight dynamics, really, but one of nervous communication. To communicate anything, a message must be sent from a sender to a receiver over a distance. For the sender to know that the message has been received, the receiver must send a message back: All's well with me; how's with you?

Such a process takes time and space—communication is a space/time coordination, a dance of rhythm and movement. Humans have developed elegant nervous systems to handle the intricacies of this coordination; the merest arching of an eyebrow or flick of the wrist requires an orchestra of neurons talking to each other as well as to the muscle groups that execute these movements.

But back to the bee and its physics, which are really not difficult to determine. (Deakin, p. 1003.) According to such considerations as air density, the weight of the bee, wing area, force of gravity, and frequency with which the bee bats its wings against the air, it is possible to determine that the frequency required for flight increases according to the square root of the bumblebee's weight and decreases with its wing area. In short, the larger the wing area and the lighter the weight, the less the insect needs to move its wings to fly. Picture the Gossamer condor, weighing less than 70 pounds, soaring without batting a wing, its wingspread wider than a DC-9. (Einstein, p. 193.)

Well and good for the condor, but the number required by the bumblebee turns out to be about 250 beats per second, a rate that the bee's own nervous physiology makes impossible to realize. Not even in humans are nerves capable of delivering electrical signals at much more than 200 impulses per second. The bee can respond only to a maximum 35 impulses per second; clearly, if the bee depended on its limited brain and nervous system, it would never get off the ground. But the bee does not know that it cannot fly. Nor does it know that automatically, without nerves, it changes the rate of its wing vibration according to how much of a load is placed on its wings. When the wind blows, the bee, with its non-valiant heart mildly beating, increases the beat accordingly, giving it no mind. In other words, the wing responds directly to the wind, and to the wing load, without the bee's mind or nervous system registering. This phenomenon is called asynchronous flight muscle response.

But you are not a bee, nor do you respond directly to the stress put on your arms and legs, but according to how fast a nervous signal travels from brain to limb.

A Wild Ride and a Quantum Baseball Game

To grasp the point about how nerves communicate, imagine speeding along with me at 100 miles per hour in a sportscar. It happens that your nerve pulses travel at around the same speed—110 miles per hour. Suddenly, just 100 yards up ahead (the length of a football field), you see a small animal dart into the road and freeze in the beam of your headlights. You see it, but can you do anything about it?

For your brain to send a message to your foot takes about .75 second, during which time the car rolls 104 feet. Only now do you begin to push down on the brake, causing the car to decelerate. With typical brakes, your car can slow down only 13 mph per second and you keep the brake pedal to the floor. How fast you will be able to bring the car to a halt depends on how fast you were going in the first place. At 100 mph, you need over 7 seconds to screech to a halt. Even while slowing, the car rolls an agonizing 555 feet farther before it comes to rest. Altogether, 659 feet of roadlength, more than *two* football fields, have been "consumed" by your 100-mph vehicle.

If you were traveling 50 mph, you would roll only 52 feet before your brain sent a message and it was received by your foot. Your car would then decelerate to a complete stop, as you stepped on the brake, in only 3.8 seconds—resulting in a complete stop in just under 139 additional feet, a total roll, altogether, of 191 feet—enough to avoid an accident.

Let us turn to another example of neural communications, this time from the sports world. A good baseball player hits around .280 or better, meaning he makes a base hit 280 times or better out of 1,000 times up to bat. Assuming no flukes, we can estimate that his response time is close to that of the flight of the baseball itself.

Suppose he batted .500: This would mean that about half the time at bat he hit the ball when he wanted to. If, on the other hand, he batted .000, he never hit the ball when he wanted to. Whatever the case, his ability depends on how long he can watch the ball from the time the pitcher throws it until it crosses home plate.

A pitcher's fastball travels at around 90 miles per hour from the mound to the batter's box, a distance some 54 feet from the position of the ball when the pitcher releases it. (Brancazio, p. 251.) The ball takes only 400 milliseconds (just under .5 second) to travel the distance. We can estimate, therefore, that a signal travels from the brain to the arms in (or just under) .5 second, less than the response time of the average car driver (.71 second).

How, then, does the batter do it? Since he cannot respond any quicker (assuming his arms' response time is about the same as for his legs), we must assume that he anticipates the ball's arrival over the plate and begins his swing well before he knows it!

Perhaps we are describing quantum physical baseball, in that, as I explained in my previous book *Star Wave* (Chapter 7), do tune in to possible futures. The ball player's batting average reflects his quantum probability of reaching a successful future. In fact, *most* sports events depend on the player's anticipating the future in order to maximize the outcome of the present.

That we anticipate the future every time we do anything in the present has been documented in experiments performed by Benjamin Libet and Bertram Feinstein at Mount Zion Hospital in San Francisco. Their studies show that a subject's brain already begins to generate waves and signals a full 1.5 seconds before the subject engages in the simple act of lifting a finger.

If quantum physics does have anything to do with human consciousness, and I believe that it does, then even events as innocuous as lifting a finger or smelling a rose must depend on the mind's ability to anticipate the experience. That future experience "shakes hands across time" with the present experience. Then quantum waves in the brain, moving from the present moment to the future, interact with mirror-image quantum waves coming from the future. The "clash" of future and present waves "creates" the present facts of life that we all enjoy or curse. Indeed, we all play in the universal quantum baseball game—the game of life.

CHAPTER 24

Give Me Autonomy or Give Me Death

The bumblebee responds to drag (or load) by beating its wings faster than its mind (nervous system) can tell it to do so. In a similar manner, each of us has an autonomic nervous system that, without willful control, processes data and causes organs to respond, just like a machine. Our body temperatures are regulated indeed, autonomically, so too our heartbeats, our lung-diaphragm movements—every function of the body.

Evolution and Autonomous Body Functions (Heart, Lungs, and Other Organs)

Why is the autonomic system necessary? Why, for example, can we both regulate our breathing if we wish, and let it take care of itself autonomically?

One reason we possess autonomicity is evolution. As we evolve, we become more aware of our autonomicity. For example, it was advantageous, when frightened, to secrete adrenalin, and it certainly was easier to allow the autonomic processes of penile erection and vaginal lubrication to occur naturally than to think about them!

Thinking about such bodily processes as adrenalin release or sexual arousal—becoming aware of them—undoubtedly was a later evolutionary development. Autonomicity and conscious will must have evolved and been differentiated from a single process. Our earliest ancestors were able, certainly, to become sexually aroused, but they did not know it consciously. Once they became conscious of their actions, a mind split occurred. Then they knew both that they were aroused and that they were able to alter the process. Each was autonomic, yet willfully aware of the consequences. Before that evolutionary jump, they virtually had no choices, and it is this ability to make choices that I am most concerned with here. Remember, a major point of this book is that through the observer effect, the act of choosing creates reality. The more choices we have, the more evolved we become. It can be imagined, too, that the appearance of moral and tribal laws took place sometime after autonomicity and conscious will were differentiated.

With the advent of biofeedback, it is conceivable that more control over autonomic functions will be achieved, and it is already known that control of blood pressure and heart rate through biofeedback and meditation produces healthful benefits.

The CNS and the ANS

In order to understand the autonomic processes of the body, we need to look at the nervous system. The human nervous system consists of two parts coexisting side by side: the *central nervous system* (CNS), and the *autonomic nervous system* (ANS).

The CNS consists of the brain, spinal cord, and peripheral nerves that provide a network of motor and sensory nerve fibers. The ANS consists entirely of peripheral nerves, sharing the same pathways as the CNS but existing separately.

Peripheral nerves are involved in the transmission of sensory information (sight, sounds, smells) to the brain and spinal cord. Nerve fibers that transmit information *from* the sensing organ to the brain or spinal cord are called *afferent* nerves; those that transmit information from the brain and spinal cord *to* the various organs, muscles, and glands are referred to as *efferent* nerves.

The ANS is separate from the CNS. The autonomic consciousness, governed by the sympathetic nervous system, consists of a series of ganglia connected together by intervening cords extending from the base of the skull all the way down to the coccyx (tailbone). It controls the movements of internal and external organs, such as heartbeat, bronchi constriction and dilation in the lungs, pupil dilation and constriction, as well as glands that are automatic in their functioning, such as those of the gastrointestinal system that secrete digestive juices, the andrenals, and the thyroid gland.

The ANS is also called, simply, the sympathetic nervous system. (Gray, p. 798.) It can be divided into what are known as the *great plexuses of the sympathetic*, large aggregations of nerves and ganglia in the chest, abdominal, and pelvic regions of the body. They are the *cardiac*, *solar*, and *pelvic* plexuses. Pelvic plexus supplies either the spermatic or vaginal/uterine plexus, depending on the sex of the person.

204

As humans evolved, it was important to develop two modes of consciousness: the willful consciousness, and the autonomic. Both overlap in many functions.

A typical example is breathing: Male humans, as I mentioned in Part Five, at rest, breathe about 12 times per minute, while females take about 20 breaths per minute, and infants take around 60. Breathing, left to its own devices, is devisive. Its automatic quality depends on the pH (degree of acidity) of the respiratory center of the brain. A change in that factor quickens or slows breathing rate.

But then consciousness enters the picture and we can, if we so wish, breathe faster or hold our breath. Similarly, using biofeedback it is possible to alter our body temperatures, our heartbeats, and some yogis even can alter the neural firing rate of their nerve cells. Indeed, there seems no clear limit between autonomic and willful control of the body, and these same spiritual teachers use the overlap of the autonomic and the willful in breathing to make the body-mind become more conscious of itself.

Quantum Consciousness: How a Yogi Controls His Heartbeat

Suppose that you had to be consciously aware of your every breath, every heartbeat, even every step of peristaltic action. Clearly, you would not have much of a mind left over for any action requiring novelty. Your attention would be so taken up with watching over these autonomic functions that you would not be able to so much as take note of a changing red light signal or to avoid the simplest accident. To enable your mind to be free from these basic functions, and yet able to take over when necessary, requires a subtle interaction between the mind and the brain.

The controlling structure for autonomic functions, located in the brain stem, is called the *reticular activating system* (RAS). The brain stem refers to a collection of nervous structures found at the base of the brain; the RAS is a diffuse collection of neurons found within the core of that stem.

Here, in the interaction between the two nervous systems, is where quantum physics suggests an answer as to how consciousness can become autonomic. If we imagine that a nerve signal from an organ, say the heart, passes along the nerve to the brain stem, a return signal will "beam" from the brain stem back to the heart. This return signal will act as a regulator, causing the heart to beat faster or slow down, for example. Normally we are not conscious of this exchange of signals; they are autonomic. But yogis have shown us that it is possible to become conscious of this exchange.

To grasp how this consciousness occurs, we first must think of the action as being conscious of itself at a very primitive level, say at the level of a single neuron. The RAS neuron "knows" of the signal from the heart and sends a signal back to the heart muscle. By "know" I mean that an act of consciousness has occurred in which a quantum wave function associated with the incoming signal has been altered suddenly by the observer effect. I call this action "popping the qwiff."

Popping a qwiff is my visualization of or metaphor for the observer effect. Quantum physics postulates that the material world simply cannot exist as pure materiality without qwiffs. Qwiff is my acronym for "quantum wave function." A qwiff represents a region, or limit, of human understanding of physical experience; it expresses a border between the hard materiality of the physical atomic elements and the mind itself. In the mathematical language of quantum physics, a qwiff has a wavelike pattern which exists as the basis for all temporal-spatial phenomena.

Each point on a qwiff surface represents a possible observation of an event, such as the discovery of an atomic particle. This wave surface moves through space and time in a logical, continuous manner. It persists in this manner until an observation takes place, at which time the wave suddenly vanishes for an instant. In its place is a single particle, or the cognition of a single event. As soon as the observation is over (and it only lasts for an instant), a new qwiff emerges, again spreading through space and time and seeking, at will, the next observation. Each observation collapses the qwiff to a single point, much as a soap bubble collapses when it is pricked. It is for this reason that I call this observer effect popping the qwiff.

If the qwiff is not popped, no return signal is generated from the brain, and the heart receives no message to slow down or speed up. The problem for the neuron is simply to note, or fail to note, the incoming signal. By the act of noting, the neuron alters the qwiff suddenly and irreversibly. This change probably occurs through the interaction of enzymes within the neuron, as well as the protein-gate structures that are embedded within the neuron's membrane and associated with neuron firing.

The yogi has probably developed a very unusual RAS system of neurons, and can actually "refuse" to take notice of messages coming from the heart. This altered consciousness may be the key to this remarkable talent—the yogi actually puts his mind into an altered state of awareness, which probably is quite difficult to achieve. Under its sway, the RAS neuron remains in a state of consciousness in which it both has and has not fired at the same time—a kind of quantum paradox. The yogi's mind has learned to take note of a situation in which both possibilities occur simultaneously, thus producing a different physical state that may resemble deep sleep, hibernation, even death. This new state of awareness cannot be achieved without focusing awareness onto something else, usually an image, a distant shore, perhaps, or some other peaceful scene. After much practice, the image isn't even necessary, just willing the non-note state is enough.

Inside the yogi's brain must lie a secret observer—a field of consciousness that chooses. The same observer lies inside each of our brains and, in fact, in every neuron of the body. It is this observer that chooses to take note, or not, of the possible events bombarding our nervous systems every nanosecond (billionth of a second).

Uncertainty always is associated with the actions of this tiny quantum physical observer. The result of observing action is to produce a change in the state of the neuron that reflects this uncertainty. For example, the neural enzyme may be a long, complex protein molecule, terminated by a molecular group than can assume either of two possible physical configurations, a or b, if noted, but that will assume an energetically favorable configuration that is neither a nor b if not noted. In the language of quantum physics, the unnoted state not a or b is a superimposition of a and b. Energetically favorable means that no action is taken, because the molecule is in the lowest energy state possible.

Let us write these possibilities as:

$$a \text{ and } b$$

$$a \text{ or } b$$

With the first, a *and* b, consciousness has not popped the qwiff, and a signal is not sent from the neuron. With the second, a *or* b, consciousness has popped the qwiff, and either the a configuration or the b configuration has occurred, with the result that in either case, a signal has been sent back from the RAS to the heart. Thus, indeterminacy still rears its head, but its consequences are surprising in that they produce an automatic or autonomic effect, the back flash of a nervous signal.

Once "trained" in this manner, the neuron continues to notice itself in the mixed a *or* b state, with the result that, after several such occurrences, the neuron is in either state, with a 50–50 probability. Surprisingly, quantum physics tells us that the same result would hold if some outside observer were to look at the *a and b* state, attempting to locate the configuration of the enzyme. The act of looking would pop *a and b*, resulting in the *a or b* state and the same 50–50 chance of finding either.

When the yogi alternates between the normal observation of the supermind overlooking the *a or b* contents of the miniminds and the supermind entering the relaxed *a and b* states that he creates, his supermind actually is taking control of the heartbeat. What is taking place is remarkable at the quantum level, but it is, nevertheless, difficult to observe. (Observation of any single neuron results in the *a* state or the *b* state; the *a and b* state is unobservable.)

The neurons are consciously changing from *a and b* states to *a or b*. In the latter, the miniminds are at work and the system is working autonomically. When the yogi relaxes or "unnotes" these mixed *a or b* states into the "pure" *a and b*, the neurons no longer have a stable firing pattern and the heart slows down, stops, possibly fibrillates. (A good yogi's heart probably doesn't fibrillate.)

Aside from the yogic ability to control the heart, we all have our hearts beating autonomically with millions (perhaps billions) of tiny neural miniminds taking note of all of the *a or b* states. Each noting of the *a or b* results in either the *a* state or the *b*. This is autonomic functioning and the body quantum in action. The development of autonomic functioning, which I take as the development of a supermind and miniminds, was a necessary evolutionary step.

CHAPTER 25

The Mind-Body Interaction

How Qwiffs Recognize Themselves as Resonances in DNA

In this chapter I offer some highly speculative thoughts about consciousness, and there are several I wish to put forward. The first concerns the behavior of qwiffs. Remember, qwiff/quantum wave functions are the quantum physicist's way of dealing with the solidity of the material universe at the level of atoms and molecules.

Whenever atoms are arranged in a highly repetitive pattern, such as those found in crystals or in the long strands of molecular DNA, the qwiffs also take on a similar pattern. This pattern constitutes a continual kind of observation in which the qwiff, in a sense, is observing itself over and over again. Quantum waves and qwiffs can be imagined as constrained by such a pattern, which, in fact, gives the structure its stability.

The qwiff, in my view, turns on and off through the observer effect. When an observations occurs, the qwiff "pops," and a point-like atom, or part of an atom, is manifested for an instant. When no observation takes place, the qwiff "hangs around," like a ghost, in the same locale in which it first popped. This sequence is highly reinforced by the repeating structure.

To try to imagine this concept is difficult because there are many atoms involved. The qwiffs, as I imagine it, are "resonating" with the structure of the molecules, so that each qwiff turns on and off with many oscillations. From the solid molecule's point of view, this corresponds to its own self-observation.

This viewpoint can be contrasted with a single atom's self-observation: It, too, can be thought of as being in a self-observation pattern, wherein its qwiff turns on and off. But being an isolated atom means that the pattern will display a higher degree of randomness. At the atomic level, this pattern appears as the atom itself, vanishing and reappearing in a sequence of random points, blurring, more or less, into a solid object.

Thus, each qwiff pattern is highly specific to the element it represents. A qwiff for the hydrogen atom is quite different in detail from the qwiff of a carbon atom.

When a sugar-phosphate molecule repeats itself as an endless chain of snakelike strands, winding around each other much like a spiral staircase, an infinite hall-of-mirrors effect manifests itself, allowing the living, conscious molecule to appear. I am describing, of course, the molecule of genetic life, deoxyribonucleic acid, or DNA.

The second idea is even stranger and more speculative. There are actually two qwiffs involved in a qwiff pop, the second of which, the star qwiff or *star wave* (as I referred to it in my previous book), is similar in form to the ordinary qwiff, only orientated backward in space-time. Thus, an ordinary qwiff, W, moving from here-now to there-then, is met by a star qwiff, W^*, from the there-then moving toward the here-now. These qwiffs multiply together, yielding the product W^*W; that is, W^* multiplying W. Now, it is not speculation that one must multiply ordinary qwiff W by its star qwiff W^* in order to calculate the relative probability that qwiff events will occur; that is exactly what quantum physicists do when they determine the likelihood that any event will occur. The speculation surrounds the idea that W^* comes from the future, traveling backward through time, much like the wave that, bouncing off the shore, travels back toward the source of the wave. I can't justify this idea by any physical experiment, at least not yet.

I believe this idea is important because it could explain how the evolution of anything can take place. My idea is similar to those that Sir Fred Hoyle discusses in his book *The Intelligent Universe*. Merely left to the odds, it is extremely unlikely that anything as orderly as a human being would arise at all simply from random processes. As I explained in Chapter 11, there needs to be some form of intelligence involved. But the question is, How does that intelligence act? Of course, I could just postulate that there *is* a Supreme Intelligence and that that being can act in any way that it sees fit. As Niels Bohr once remarked to Albert Einstein, when he was trying to figure out how God did it, "Stop telling God what to do."

I certainly don't want to do that! But I do want to know how God does it. Yet, as a physicist, I am somewhat constrained: I can't postulate just any idea, because a scientific idea, in order to be considered valid, must fit with what we already know (or, at least, "think" we know). The idea that W^* comes from the future may just save the day, however. As Hoyle puts it:

If events could operate not only from past to future, but also from future to past, the seemingly intractable problem of quantum uncertainty could be solved. Instead of living matter becoming more and more disorganized, it could react to quantum signals from the future—the information necessary for the development of life. Instead of the Universe committed to increasing disorder and decay, the opposite could then be true. (Hoyle, p. 213.)

In a highly organized material containing repeating patterns, the W*W content becomes highly repetitive, producing a probability pattern of reinforced strength. Thus, crystals of repeating materials, such as sodium chloride, carbon lattices (such as diamond), and other single crystals of metals and metals in combinations with other substances, possess great strength or other unusual properties.

In DNA we have a similar phenomenon of great repetition, with complex patterns of sugar-phosphate backbones interrupted by the much longer, seemingly random steps of *base pairs* linked together in complementary codes. These bases, you'll recall, occur in four types: A, C, G, T.

Here a third idea surfaces: Because of the repetition of the DNA structure, the likelihood of a repeating W*W pattern is highly enhanced, with the W* involved propagating from a near future to the present. The signal from the future is more or less the same as that from the past, and the pattern, consequently, tends toward stability. The more stable the pattern, the less likely that the distant future will disturb it. Again, the idea that there exists a resonance between the qwiff and its structure—involving both the past and the future—is at play here. Signals from the distant future do arrive, however; without them, DNA would never alter its patterns. But the more stable the reinforcement brought on by the repetition of the strand, the smaller the disturbance produced.

It is the interplay of the endless crystallike repetition of the DNA strands, twisting in space and dancing in time as vibrations with the almost though not quite random patterns of A, C, G, and T bases, that produces stable animal and plant consciousness. Consciousness, as we commonly experience it, thereby emerges as a consequence of the qwiff vibration patterns associated with DNA vibration patterns repeating and resonating with both the future and the past.

Molecules of DNA within shouting distance of each other also vibrate, sending quantum semaphore messages back and forth, and in this manner a resonance arises between neighboring molecules. This resonance is much like any other resonance phenomenon, such as a building's vibrations in the wind or the rolling of a ship on the high seas. With energy being fed from one molecule to the other at just the right frequency to induce the other molecule to respond, the two resonate together. It is this resonance of waves in different cells that could result in the healing of the cells.

Illness could result from an opposite effect. When molecules are off-resonance, they fail to communicate with each other; such off-resonance could arise from atomic changes in the molecules or from subtle changes in the probability patterns of the qwiffs, possibly brought on by negative thinking. Influenced by such thinking, perhaps molecules tend to isolate themselves, forming self-contained units of limited capacity. Such molecular isolation can be understood in terms of our own behavior when we feel depressed or unduly anxious about something, and want to be alone in our misery.

Consequently, illness and negative thinking could create molecular islands of separation within our cells. Healing energy counters this separation tendency by fostering correlations between molecules: One molecule heals another.

And in the relationship between a healer and a healee, the healer attempts, through touch and simple bodily presence, to resonate with the healee. Healing energy is felt, then, as a simultaneous presence in the healer and the healee.

Feelings and Energy States

According to modern physics, all energy is quantized: Energy moves from one place to another in whole, observable units called *quanta*. It also is true that energy is measurable in whole units, but the Heisenberg principle of uncertainty sets limits on how accurately any energy can be measured or known. This limiting factor is time, and knowledge of energy becomes a question of timing.

When atoms or molecules emit or absorb energy, and thus undergo energy transitions, the time interval during which that energy transition occurs must remain uncertain. If, on the other hand, the time interval is known, the energy change becomes uncertain. Since we now know that energy and frequency are one and the same, this means that the frequency change and the time when a frequency change occurs are not both knowable with certainty: The more you know the frequency, the less you know the time, and vice versa.

This complementarity between time and frequency (or energy) is analogous to what happens when you sing a song. To know that a certain musical note or frequency is being sung, the singer must sing the note for a long enough time in order to allow many complete cycles of vibration to enter the ear. If he or she allows a wide enough berth for the time interval—that is, does not specify the exact length of time that he sings the note—he can tune in to the frequency-energy of the note with a high degree of accuracy. On the other hand, if he attempts to sing the note too quickly, so that the complete note cycle is not sung, the exact frequency-energy of the note cannot be determined.

Now according to my star wave theory, energy is feeling. Energy and feeling, which then arise as an emotional state, are related in the same linear way that energy and frequency relate. Associated with a change in energy state is a corresponding change in feeling, which tends to produce an emotional state in the observer. Thus, all energy transformations in the body ultimately are felt as emotional-state transformations. Also according to the star wave theory, time and thought are related linearly: Thought becomes known at the expense of feeling in the same way that time is known at the expense of energy. If thoughts are carried out while a body is undergoing energy transformation, therefore, the energy transformations, or feelings, are rendered indeterminable. This principle is extremely important in understanding how the body, through thought, can be made ill or can be healed.

This principle also sheds light on the energy-transformation process I celebrated in Part Two: "Eating." Most of us eat food unaware of the energy transformations and bodily feelings that are taking place. This is because most of us eat with thought always present; we are either talking to someone else between bites or, when we eat alone, thinking about the world and its influence on us. Few of us take the time to become uncertain about time and thought, and we eat according to schedule. Food may not undergo complete or proper energy transformation, consequently, and we then have indigestion. In fact, indigestion is rampant in Western society.

By tuning in to the food we eat, we can rid ourselves completely of indigestion. Tuning in to eating means becoming as aware as possible of the taste of food as we chew it. It means focusing on the smell of food, on its texture and temperature, on its degree of acidity or alkalinity. It means noticing how much or how little we salivate as we eat, dwelling, in short, on the totality of the eating experience. In so doing, it also becomes impossible to overeat. The body, unencumbered by thoughts, becomes highly sensitive to energy transformation. Foods taste better when eaten with an empty head as well as an empty stomach!

They also provide maximal nutrition. The simple action of becoming aware of food, and not thinking or talking while we eat, provides for better salivation and, thereby, better digestion. Undigested or poorly digested food means poorer nutrition. Conscious eating clearly causes the body to make better use of the food entering into the cellular processes of life. Eating at its best is a conscious business.

Another example of the star wave principle occurs in listening. Most of us rarely hear the essence of words spoken by others. What usually happens instead is that our own thoughts interfere by taking the speaker's words and placing them in an overlay personal context. This mistaken context often results in confusion. For example, the word "refuse" could jar the memory, returning the listener to a past experience that was particularly uncomfortable. Maybe the listener was refused entrance to a home or theater at one time, and when heard again, that word may strike a chord, a fear of being refused entrance once again. Similarly, the word "love" is often misunderstood. If I hear the phrase "I love," I can easily project a meaning onto it completely different from what the speaker has in mind.

At the level of quantum physical DNA encoding, two messages are being superimposed: the words of the speaker, and the thoughts of the listener. Since the speaker's words cause energy transformations in the body which are experienced as changes in emotional state, the listener's words can do no less than alter and change the meaning of those words. Indeed, the early childhood programming we all underwent was altered drastically when we learned to speak. Our childlike words became attached to mother or father's spoken words. In this manner, a child can learn either to love or to hate an instruction provided by a parent.

For example, suppose the parent is teaching the child appreciation of food. The child, in his or her high chair, is told over and over again, "Oh!, It's so *good*! Yummy, Yummy," even though the food in question tastes yucky to the child. The word "good" is now associated with force feeding. As a result, the child may end up with a mistrust of *goodness*. Later in life, when something is too good, he may become suspicious. If programmed too severely by parents, one even may equate "good" with "I hate it."

In a similar manner, we all have become resistant to certain kinds of new information that does not fit our earlier programming. Words or thoughts race into our minds, disrupting the new information patterns that otherwise could arise. Thus, for every miracle, a thought of skepticism; for every logical thought, an irrational feeling.

Brainwashing attempts to tire the mind's chatter by an endless repetition of new data; eventually, the muscles and energy transformations involved in the formulating of thoughts are depleted. The new information then is allowed direct access to DNA molecules, and a new personality, or a brainwashed mind, is born.

The early disasters with consciousness-raising religious movements was misunderstood by the general public. Attempts to find lost children, root them out of the movements, and reprogram them to fit in to normal society through rational-emotional conquests by a reprogrammer did meet with success. But what was not understood was the quantum physics of all mental programming. In a sense, every individual is brainwashed as a result of the simple occurrence of birth into a given society. Therefore, the reprogramming of young people by consciousness-raising religious groups is nothing more than the application of repeated messages uninterrupted by the thoughts of the listener. Similarly, during times of war our presidents are able to assume father-figure-like proportions, telling us to march off to battle and lay down our lives. Our presidents are not brainwashing us, they simply are reinforcing old programs. This is the major problem with all political leaders: They attempt to tune in to the consciousness of their constituents but usually are resonating with the past views of those same constituents. With the increase in communication brought on by the computer-technology revolution, perhaps they can tune in more successfully. The Vietnam War might have ended much sooner if the computer revolution had taken place a little sooner.

There is, of course, a positive side to brainwashing. We are never too old to be reprogrammed. It is possible to change our thinking patterns and widen our horizons, surpassing the old programs and even becoming spiritually freer and happier people. The so-called mid-life crisis is just such an opportunity, and in the 1970s, many consciousness movements did just that, with much success: They altered the consciousness of many and produced a consciousness revolution.

Health, understood in terms of the facility offered through quantum physical understanding, becomes a question of proper attention. By using thought, old programs can be annihilated. DNA can be altered to produce a different body. In a similar manner, the repression created by the transformation of thought into anger can alter DNA and could lead to cancer.

Other injuries also arise as the result of thought transformation. Louise L. Hay, in her book *Heal Your Body*, gives a long list of how negative thoughts become targeted as body diseases. For example, cancer is a deep secret or a long standing resentment or grief eating away at the self; constipation arises from refusing to let go of old ideas; coughs arise from irritating thoughts not spoken; deafness occurs from a desire not to listen; and so forth. Certainly constipation can also be caused by a low-fiber diet, and cancer can result from contact with carcinogens. Our physical environment can produce many hazards for life. This we will explore further in the next part of the book on healing.

How Thoughts Become Physical

How can a negative or a positive thought become the body? In order to grasp this idea, we need to look at the momentum and position of objects in space.

The location of an object in space is complementary to its momentum—how fast the object is moving and with how much mass. The momentum of an object is the product of its mass and velocity. In quantum physics, momentum is a primary quality; mass and velocity are secondary. Thus, an object can have a well-defined momentum but not necessarily have a well-defined velocity or mass.

Complementary to momentum is spatial location. To know where something is, is to lose information about how it moves; motion implies a sense of direction. For example, if you know a molecule is located at x, where x is the distance from a given plane in space, you forfeit knowledge about its momentum in the x direction: You don't know whether the molecule is moving away from or toward the original plane. This, again, is the *uncertainty principle*.

According to quantum physics, momentum is related to the wavelength of the qwiff. Actually, momentum has the same relationship to the wavelength that energy has to the qwiff's period of vibration. Dividing the number 1 by the vibration period yields the frequency of the qwiff; dividing 1 by the qwiff's wavelength produces a quantity called the space frequency. Thus, momentum is the same thing as space frequency. Energy is the same thing as time frequency. Just as energy can be given as $E = hf$, momentum is $p = hk$, where h is Planck's constant of action-uncertainty, f, the time frequency, and k, the space frequency.

A simple analogy exhibits the uncertainty principle applied to location and momentum. Think of yourself as a child who has become lost in a city. Imagine that you have come to a city block that stretches miles and miles ahead, with no end in sight. As you run along the sidewalk of this block, you notice that you can see no houses, and, instead, there is a ten-foot-tall picket fence running along the sidewalk.

As you move along the street, you observe that you are passing one identical fence stake after another. Since stakes are arranged periodically, they are spaced at equal distances. Now, think of the picket fence as a state of well-defined space frequency, with an associated well-defined momentum. You are attempting to locate where you are along the fence, but, looking in either direction, no end is in sight. You do not have a well-defined location in space.

After a while, you see a broken stake, a disruption in the pattern of space frequency. This break is a clue as to where you are; that is, in order to locate yourself in space, you must break the constant pattern of space frequency. By giving up a well-defined momentum, you provide a well-defined position.

In the opposite manner, signs or breaks in patterns exhibit spatial location, but no regularity—no insight or direction into the future. To know where you are is to give up knowledge of where you have been and where you will be. To know where you have been, where you are, and where you are going—to know your flow pattern—is to give up any sense of where you are now.

I often feel that Buddhist monasteries, because of their emptiness, provide, by design, the timeless sense of flow of no-time–no-space. The flow of Buddhism appears to me the same, always the same, whether it occurs in this century, the future, or a thousand years in the past.

In *Star Wave* I offered the theory that momentum and intuition are the same in the brain, and that location in space and body sensation were the same. Thus, external events arise as bodily sensations. I see a star and my retina undergoes a transformation that is sensed; hearing a song, my eardrums vibrate.

We again begin to see here how a negative or a positive thought could become the body. Through our conscious choices, which depend on how we choose to look at the data of the outside world, we ultimately alter the momentums or locations of physical objects in space. These objects exist in our bodies as molecules and atomic structures, and they change in order to register our outside experiences. In effect, they correspond to representations of outside-world experiences, and any one can be taken in any number of ways, according to the momentum-position choices we make in our inside worlds—our own bodies.

Momentum appears as continuous, steady movement. Without realizing it, we tune to momentum with our intuitions. We see a football fly through the air, and, while it is flying, we know where it will land. We watch the endless flow of the river, sense its direction of flow, and know where it will go next after it passes by us. In this way, we use our intuition to internally process momentum experienced in the outside world.

It is in this way that external experiences, words, and physical forces are experienced bodily; they end up as interplay between intuition and sensation, and, for that reason, tend to localize at key body sites. A disruption in intuition surfaces as a sensation felt in the body. We feel hurt when our spouse says "Good morning" in the wrong way; the normal flow of countless good mornings is disrupted by an inattentive greeting, and our bodies undergo wincing pain.

As children we learned from our mothers. With her acceptance and flowing, loving touches, our skin opened to more sensations. Full breathing and respiration took place. The nerve endings surrounding the pores of our skin made them bristle with anticipation of her loving touch. If, however, she became distracted and suddenly started to handle our baby body roughly, our skin recoiled with fear, our pores closed up, we did not want to be touched. This could lead to several setbacks in our later life. As teenagers and adults, our skin could take on these wishes, and we could develop acne, shingles, even psoriasis.

Through life's experiences, every experience leaves its mark as a body sensation caused by interrupted intuition or flow pattern. Just where bodily sensation is experienced depends on several possibilities, which, in turn, depend on the pattern of experiences being felt. In this manner we learn.

Learning moves through our bodies as a dance of intuition, with body sensations matching exactly the flow of momentum and the location of external events. In our bodies are stored patterns that are manifested as body posture, muscle tone, blood flow, and slight alterations in water pressure in our cells. These physical variations, in turn, gently alter the cellular processes, reaching all the way down into the cellular nucleus where the pool of DNA buffeted by water molecules undergoes changes in vibration. The vibrational dance of DNA changes, and, depending on the nature of the external experience, forms a stable pattern, matching the movement of the external momentum-sensation interchange. Once the pattern is established, it becomes "us."

PART SEVEN

Healing

heal To restore to health; cure. To set right; amend. To rid of sin, anxiety, or the like; restore. To become whole and sound; return to health.
—Adapted from *The American Heritage Dictionary of the English Language*

The root word for "healing" comes from the Indo-European word *kailo*, which means "whole," "uninjured," "of good omen," and also "holy." Thus, to be healed is to be "wholed," at one with the universe. We all respond to a loving hand, a sympathetic heart. Sympathy is a rhythmic understanding—a vibration of two together in harmony and phase. A sympathetic person can "feel" what another person is feeling.

What happens when someone undergoes healing? Although we all can see a wound heal, or are aware of a broken heart on the mend, particularly if that heart happens to be our own, no one knows just what constitutes the healing mechanism. Here again the body quantum comes to the rescue. I speculate that healing is a quantum physical process that can be understood as a phase harmony of quantum waves. By being with each other, we enhance this harmony, just as two magnets provide more strength than each magnet separately. To be healed, separate parts of the body also must begin to respond to each other in phase harmony.

Accordingly, illness is defined as vibrations of quantum waves out of harmony with each other. Using quantum physics as a basis, diseases, in addition, are seen as a game of probability modified by our own deepest thoughts and wishes. We become ill, we feel depressed. But which came first, the illness or the depression? In quantum physics, the answer is not so simple. For the depression and the illness, in a certain sense, are the *same thing*. The connection between a physical disorder and a mental thought is very subtle but consequential: Disease leads to depressive thoughts, depressive thoughts lead to disease.

The physical body and the mental thought are correlated. Illness, then, is seen as a choice, perhaps unconscious, made by the invisible observer inside us all. And these choices are governed by several factors, including the physical and mental condition of the body and mind.

Along with the discovery of the importance of quantum physics in the body comes a new realization about diseases: Some resist all attempts to be cured, and these are *quantum mechanical diseases*.

CHAPTER 26

The Quantum Physical Diseases

A Vacation and a Flirtation with Quantum Physical Diseases

I recently returned from a long vacation, during which I completely altered my habits of daily exercise, consisting at that time of either jogging or working out with Nautilus equipment. I also ate more, slept and lounged more, drank, smoked. I read easy and light novels, and failed to write a single word for more than a month. I had a marvelous time! But I came back feeling out of shape, groggy, and a little guilty for having abused my body so much (even though I'd enjoyed it!).

The questions that nagged at me when I returned, I'm sure, were the same as those that nag us all: What harm did I do to my body? Does overeating, alcohol consumption, and smoking really take years off my life? What about lack of exercise or mental stimulation? Was my vacation more costly than the dollars I paid? Did my one or two cigarettes a day really increase my chances for getting cancer or emphysema? Does drinking alcohol *cause* liver cirrhosis? Did that one month reprieve from my normal full physical program of running and Nautilus set me back so far that I will never be able to recover my previous degree of fitness?

On the other hand: Could that rest have added years to my life? I really enjoyed the vacation and the reprieve from discipline. But what effect do rest and indulgence have on health and vitality? It appears that we all need both, at least every once in a while, if we are to remain happy.

When I finally resumed my more temperate life-style, I had a number of surprising insights. I realized that during my vacation my tolerance for abuse had increased with increased abuse. Usually my habit of having one cigarette a day, combined with my vigorous exercise program, produces headaches, and as a consequence, a decreased desire to smoke a second. On vacation, I was able to smoke three cigarettes a day with little aftereffect. Also, I was able to drink alcohol two or three times a day—something I can't do when I'm involved in my ordinary daily routine. It seems that when I am in shape, I have little tolerance for abuse, but when I am not, I can withstand and, indeed, even enjoy much greater abuse. No wonder alcoholics enjoy alcohol so much, and two-pack-a-day smokers enjoy that fortieth cigarette just as much as the first one in the morning.

Even though I indulged myself more on my vacation than I normally do when I am at home writing, I enjoy a stress-free existence. It seems that my mental state has more to do with the onset and accelerative rise of discomfort or disease than my simple, physically abusive behavior. In other words, although my organs were getting out of shape, my muscles and ligaments slackening, and my lungs dealing with oxygen starvation and carcinogens brought on by cigarettes and other smoking materials, my spirit seemed to be having a whale of a time.

Jogging, eating wholesome food, and writing books keeps me quite fit but also anxious, because I *worry* about keeping healthy, wealthy, and wise; and I know that prolonged worry can lead to illness. The question is, Does anything I do to prevent disease really prevent disease?

The answer is no longer simple. For one thing, the near wipeout of contagious illness, such as polio, smallpox, and tuberculosis, has changed our very concept of disease.

Our changing concepts are reflected in James F. Fries' and Lawrence M. Crapo's *Vitality And Aging*. They point out that the U.S. Department of Public Health frequently publishes statistics on the major causes of death. (Fries, p. 62.) Fourteen categories are listed that include almost every cause of death:

Infectious Disease	*Violent Death*
Diphtheria	Motor vehicles
Measles	Suicides
Pneumonia/flu	
Smallpox	*Chronic Disease*
Streptococcal	Cancer
Syphilis	Cardio/renal
Tuberculosis	Diabetes
Typhoid	
Whooping Cough	

There have been spectacular mortality declines in the first nine diseases; six are now listed officially as zero mortality, meaning that only 1 person in 200,000 dies from them in a year. These include whooping cough and smallpox, for example. The last three killers, the chronic diseases, are still a problem.

222

Our present health habits and our increased medical knowledge have nearly wiped out all infectious diseases. However, there are still obvious dangers such as cancer and heart disease. The big question is, Why do some diseases persist?

Facts about Disease, Aging, and Death

Many of us hope that medical science will wipe out all disease through medication. We believe that the human lifespan is increasing, and that death is the result of disease. Many believe that aging is controlled by some program in the brain and genes.

When looked at carefully, each of the above statements turns out to be misleading if not downright false. The human lifespan is not increasing: It has remained fixed, as far as medical and other scientific records are concerned, for 100,000 years. What has increased over the years is the average life expectancy (ALE), a number based on statistical factors. Life expectancy is the expected age at death of the average individual, granting current mortality rates from disease and accident. In the United States, the ALE is around 73 years, and rising.

Lifespan actually has not increased. The maximum life reached by a human appears to be 115 years, on careful inspection of records. Lifespan, taken as the age at which the average individual would die if not subject to disease or accident, has been 85 years for the last two centuries.

But death does not require disease or accident; people die even when free of disease and accident. And for these people, the average age of death has not changed significantly.

What has changed is awareness of disease. Today we know more, have better statistics, and are able to control many disease factors. I would like to suggest that diseases can be separated into two groups: the *classical mechanical*, curable by such modern antibacterial agents as penicillin, and the *quantum mechanical*, which are not so easily treated. These latter include, but are not limited to, viral infections, cancer, atherosclerosis, diabetes, and several others.

The quantum mechanical diseases (QMDs) are with us today as they were in earlier times. Medical treatment does not appear to be the best way to solve these current health problems. The major chronic or quantum mechanical diseases are the biggest health threats. And the best treatment today involves not medicinal but the prevention of factors that are likely to induce decreased vital capacity.

"Vital capacity" is a difficult term to define. It refers to such obvious health signs as the ability to take air into the lungs, hold your breath, and blow out a candle at a certain number of paces; the velocity with which your nerve cells conduct electrical signals; the basal-metabolic rate; kidney blood flow; and other vital body functions. As we age, these functions all exhibit declines. For example, a person 20 years old possesses around 100 percent functional use of all of these capacities; but by the age of 40, all of these functions have decreased, the greatest being in maximum breathing capacity, to about 90 percent. By age 80, breathing capacity has decreased to nearly 40 percent, with the other functions not as low (nerve-conduction velocity decreasing the least, to about 90 percent).

I believe that the decrease in vital capacity with age is connected directly to quantum physics and results from errors occurring at the level of molecular processes, such as the building of complex protein molecules. It is these errors that produce the decrease in vital capacity that we all call growing old.

Aging appears as decreased capacity for life in all of our organs, probably from birth but certainly from age 30 onward. Thus, we arrive at three facts of future life:

• The human lifespan is fixed.
• Vitality decreases with age, and, in fact, is age itself.
• Quantum physics gives life, and takes it away.

What Is a Disease?

What exactly does quantum physics have to do with disease? After all, according to the dictionary, disease is merely "an abnormal condition of an organism or part, especially as a consequence of infection, inherent weakness, or environmental stress, that impairs normal physiological functioning." However, this definition is not complete. What determines whether an organ is functioning normally? Furthermore, not all diseases are infectious or contagious. While tuberculosis can spread from one person to another, diabetes cannot. Findings indicate that some diseases, in particular those that most people die from today, are universal. It seems that nothing can be done to prevent them from occurring.

The fact is that these universal diseases, although they may lead to death, are, at the same time, *needed* for our very life. Let me clarify this rather shocking point.

I propose that the universal diseases arise out of disharmony of quantum physical probability patterns within and beyond the human body. These patterns are responsible for both the normal functioning of human organs and cells and their abnormal functioning as well. That is why I refer to these universal diseases as quantum mechanical.

In order to grow and to change, to modify one's body and mind, one must take risks. Indeed, life is a chancy affair. These risks produce both benefit and ill health. The athlete and the genius are prime examples: Few top athletes escape injury; few great minds fail to experience depression; few artists fail to feel that, at times, their art is bunk.

In other words, patterns of risk occur, associated with mental processes—often not spoken but simply felt. These patterns are quantum physical in origin and arise both from conscious thought and from conditional, unthought feelings and intuitions with regard to appropriate behavior. They occur not only in humans but also in animals. It is the inherently risky business of life that is responsible for life's ability to change, grow, and adapt itself to a variety of environments and, at the same time, inevitably to give rise to universal diseases.

I base these observations on the findings of Drs. James Fries and Lawrence Crapo. They argue, in *Vitality and Aging*, that the universal diseases are by far the greatest current health problem and that searching for cures for these diseases remains inappropriate. The diseases include: atherosclerosis, cancer, osteoarthritis, diabetes, emphysema, and cirrhosis; they are the quantum mechanical diseases.

Fries and Crapo describe another category of disease that includes Hodgkin's disease, ulcerative colitis, insulin-dependent diabetes, rheumatoid arthritis, psoriasis, multiple sclerosis, muscular dystrophy, and schizophrenia; these I label *semi-classical*. The conditions for their occurrence neither are universal, as for the quantum mechanical diseases, nor are they acute or classical mechanical, as are any of the infectious diseases, for example.

These diseases are somewhat mysterious and cannot be attributed simply to universal decline in vitality or to a "bug." Yet they appear somewhat like acute or infectious diseases; they can spontaneously disappear, and they may reverse impressively with medication or following an operation.

Arteriosclerosis also is difficult to classify in this manner because it is not universal. Arterial degeneration, referring to the loss of elasticity in the vessel walls and fibrotic change, probably is a universal or quantum mechanical disease.

The point is that the universal diseases (labeled chronic by Fries and Crapo) pose by far the greatest threat of death and, therefore, the greatest health hazard in the United States.

On the other hand, the conditions for the occurrence of classical mechanical or infectious diseases do not appear to depend on the aging process. These curable, contagious diseases I call classical mechanical diseases because they are spread by the passage of classically defined, small microscopic bodies from one individual to another. Just as classical mechanics was being replaced by quantum mechanics in the early part of the 20th century, so too the classical mechanical diseases were being wiped out. It is striking to realize that the death rate from acute infectious diseases, such as diphtheria, typhoid fever, measles, and smallpox, were as high as 40 per 100,000 in 1900, the year quantum physics was born. As quantum physics grew, these diseases declined, so that today virtually no one dies from them.

Fries and Crapo argue that we all are dying of the universal or *chronic* diseases. (Fries, p. 86.) The name of the game is not the prevention of these diseases, but a strategy for controlling the progression of organ malfunctions, which appear as the symptoms of these diseases and which are bound to occur. We all die of something.

They explain, "The myths of the super-centenarian, of Shangri-La, and of Methuselah crumble upon close examination. Yet, in the rubble, we find hope for a meaningful kind of rejuvenation—where the maximum lifespan is not prolonged but the period of vitality is."

From a quantum mechanical point of view, all diseases are stimulated continually to arise through not only the mechanisms mentioned—infection, inherent weakness, or environmental pollution—but also through the mind and in its interactions with the body. The mind-body interaction arises in the decisions we make about how we conduct our lives. Because, in my view, the mind-body interaction is quantum physical, these decisions do not always result in success. Life is a crapshoot: The daredevil may die at 30 jumping out of an airplane, but he might succumb to cancer, if he weren't a daredevil, at the age of 50.

The main difference between universal diseases and contagious diseases lies in the size of the perturbing influence. Microbes and some viruses are classical mechanical, so to speak, in their actions on the body. Drug therapies wipe them out. The quantum mechanical or *fundamental life-processing* diseases are far more insidious. They appear as the slow decline in organic functioning that we all experience in the processes of life itself. The examples of the decline in nerve-conduction velocity, the slowing of the basal-metabolic rate, the decline in blood flow to and from the kidneys, and the decrease in the maximum breathing capacity of the lungs are quantum physical, nonpreventable, and universal. Research has shown that the vital capacity associated with these functions continues to decrease linearly with increasing age, starting from about age 30.

Just as quantum physics brought out new and previously undetected physical phenomena, such as electron spin and nuclear forces, the wipeout of classical physical diseases has made the quantum diseases stand out as the killers. With ever greater sensitivity to life's processes, we become ever more aware of its limits and its finality.

CHAPTER 27

Cancer, or How a Cell Cries for Immortality

The death rate for cancer has risen steadily over the years, from 64 per 100,000 in 1900 to around 180 per 100,000 in 1980. What can we learn from such a statistic? Possibly, we may begin to think that cancer is on the rise; a little research, however, shows that this is simply not true.

How Cancer Cells Are a Cry for Life

Most of us have heard comments that cancer is becoming epidemic, that smoking causes cancer, or that eating meat introduces carcinogens into our systems. Carcinogens, we are told, are the cause of cancer. We also are told that cancer is a virus that can be wiped out, like the polio virus. These statements and others turn out to be misleading, false.

First, if we compare the tuberculosis death rate of 1900 (914 per 100,000) with that of 1980 (about 2 per 100,000), we begin to see why the death rate of cancer has risen. In 1900, if you died early from tuberculosis, you didn't have time to get cancer—TB got you first. In 1980, with TB wiped out, you now live longer and therefore are more likely to die of cancer. It isn't that cancer is on the rise, it's just that its effects as a killer were previously masked by the deaths due to such other diseases as TB.

A similar rise in the occurrence of quantum mechanical, or universal, diseases is seen over the same 80-year period.

In other words, instead of infectious diseases wiping us out, a new threat to health is emerging. This new threat requires a shift in our conventional wisdom and the recognition of the emergence of chronic, universal disease. It will not be sufficient to use the same medical techniques in dealing with diseases I call quantum mechanical. To develop a new way of thinking about chronic disease, next we need to develop an understanding of cancer.

What Is Cancer?

A surprising answer is now apparent: Cancer, paradoxically, is both a killer and life's attempt to become immortal. That this dichotomy is true became apparent in 1961, following painstaking research, when Leonard Hayflick and P. S. Moorhead published a landmark paper that demonstrated that normal human fibroblast cells (special cells that form scar tissue by manufacturing collagen and other protein substances) have a strictly limited lifespan in tissue culture. (Hayflick, p. 585.) Cells taken from normal human embryos were observed to divide about 50 times, but no more. This research has been duplicated by several investigators, and today is called the *Hayflick limit*.

What was amazing about this research was that it flew in the face of previous experimentation in 1912 by Carrell and Ebeling that showed that normal chick embryo cells, if cultured continually, would live forever. They demonstrated that cells *in vitro* (that is, in petri-dish tissue cultures) were immortal.

Certain cells are immortal. The precursors of the egg and sperm must be; if not, the species would not continue to survive. The history of any species depends on the successive dividing, over and over again, of the original seed.

Under optimal conditions, investigators have shown that some cancer cells also divide indefinitely—the human malignant cells, for instance, such as HeLa cells (the name is taken from a woman whose name was Helen La——, who died of cervical cancer). Recent research at MIT on keratinocytes (a type of skin cell) show as many as 150 cell divisions, suggesting that these human cells would, *in vitro*, continue to divide forever. (Pearson 1, p. 138.)

We all know that our skins, when cut, begin to produce new skin cells by regeneration and cell division. Similarly, intestinal-lining and bone-marrow cells continue to divide *in vivo* (on site within our bodies) as long as we live.

COMMITTED AND UNCOMMITTED CELLS

In 1975 Kirkwood and Holliday proposed an ingenious explanation of the Hayflick limit. (Fries, p. 52.) Their theory assumes that, at first, all cells are immortal. Upon division, two types of cells are produced: committed, and uncommitted. Committed cells are mortal, the uncommitted ones, immortal. With further division, uncommitted cells continue to produce both committed and uncommitted cells, but the committed cells only produce committed cells. Eventually, the culture contains few uncommitted cells compared to the large numbers of committed ones.

Although after hundreds of experiments as yet there is no evidence for the existence of any uncommitted cells, I still feel that this theory warrants further research. The question is, To what is a cell committed? I will look into this later.

Now I bring into this discussion the idea that cancer cells are uncommitted, and, therefore, capable of dividing forever. From my quantum physical perspective, a cancer cell represents a consciousness that wishes to reproduce itself because it "thinks" it is the *only* entity around. It divides, and then each offspring cell also "thinks" that it is the only cell there is—and reproduction runs rampant.

On the other hand, committed cells stop dividing because they "sense" that they are part of a bigger whole. They are in communication with other cells in their environment, probably through quantum physical boundary conditions that limit the energy-level structures available to the DNA within the cell. With each division, the committed partners are aware of the presence of the others. This begins to alter the energy-level structures so that by the time the Hayflick limit is reached, further cell division becomes energetically unfavorable.

From an entropic consideration, a single cell undergoing division must increase its surface-to-volume ratio by producing more surface than is needed to encompass the same volume. This tendency results in entropy production, usually by giving off heat to surrounding material. Thus, it appears to be entropically favorable to have cells continue to subdivide, given that all other entropic conditions remain fixed. Perhaps cancer cells too are nature's attempt to maximize entropy.

CHAPTER 28

The Cause and Cure of Quantum Physical Disease

Every year one hears claims that the cure for cancer is rapidly approaching and that the cancer cure rate is improving. We hear of many miracles and begin to wonder if the AMA somehow is holding out on all of us. According to Dr. Haydn Bush, who treats cancer patients and directs the London Regional Cancer Centre in Canada, these claims are not true. (Bush, p. 34.) Modern treatments, such as radiation therapy, chemotherapy, and surgery, only appeared to be winning the battle against cancer when those procedures and their survival statistics first appeared in the early 1940s.

This statistic, soon a gold standard of clinical oncology, was known as the five-year survival rate. This usually meant that if a patient survived five years after first diagnosis, he or she could be considered cured. The five-year survival rate, however, has proved to be misleading. Bush and his colleagues have noticed that patients who were "five-year cured" were still likely to develop cancer again later in life. This statistic was keenly felt by my own family. My father first contracted cancer at the age of 39. His diseased kidney was removed, and he was told he was cured when no further cancer was detected five years later. However, the cancer returned and he died of it at the age of 48.

The Quantum Physics of Causing and Curing Cancer

For many kinds of cancer, it doesn't appear to make much difference what kind of treatment is used. Thus, positive results suggesting cures should be taken with caution. In brief, it appears that our whole approach to cancer, particularly the most serious forms, needs to be radically changed if we are ever to cure it. Current research on the mechanisms of cancer causation and progression, and on possible anticancer substances in our diets, may lead to control through prevention and treatment.

However, I doubt that this approach will amount to anything more than a temporary stopgap—a means for keeping cancer in remission, but not an actual cure. If cancer is a quantum physical disease, and if, as many physicists are beginning to accept, quantum physics does relate to consciousness, then it follows that cancer and consciousness are related and probably correlatable. I realize that I skate on thin ice in saying this. Could this mean that a person could cause cancer cells to proliferate merely by continually thinking negatively? After all, many of us have lost loved ones to cancer, and I'm sure that we felt that they were not necessarily gloomy or depressed before contracting it.

So what causes cancer? Do depressive thoughts cause it? In other words, what are the causative factors producing cancer? It appears that a highly improbable sequence of events must occur at the molecular or quantum physical level in order to produce a cancerous cell.

Carcinogens and What They Do

It is now thought that the cancer process begins through a cell's contact with certain substances, carcinogens. Benzopyrene from cigarette smoke is a typical example of a carcinogen. However, contrary to popular belief, carcinogens in their original form do not cause cancer. It is clear, for instance, that smoking itself cannot be the cause of lung cancer or else most smokers would get it. In fact, in surveys one finds that the large majority of cigarette smokers do *not* get cancer. However, it is also clear that among the population of lung cancer sufferers, most are smokers. (It may even be true that cancer itself may a causative factor in inducing people to smoke!)

Once a carcinogen has been introduced into the lung, for example, the carcinogen must undergo a molecular or quantum mechanical transformation. This process is called *carcinogen activation*. Paradoxically, the process of carcinogen activation is used to render the invading substance harmless by detoxification, allowing the carcinogen to be converted into a form that is excreted safely through urine or sweat glands. (Did you ever notice how a smoker's body odor is stronger than a nonsmoker's? This is caused by the process of carcinogen activation.) The activation process is caused most commonly by a set of enzymes within the cell. For some as yet unknown reason, some people possess unusual combinations of detoxifying enzymes that, again for reasons unknown, perform a different modification on the carcinogen. Consequently, the carcinogens now are able to cross the cell's nuclear membrane and enter the nucleus, where they encounter the central building-block mechanisms of all cells: deoxyribonucleic acid, DNA.

The cell has its own police action available, however. Within the cell are scavenger molecules—the microimmune system—that seek out and bind themselves to activated carcinogens *before* they enter the nucleus. Even if the activated carcinogens manage to evade this system, they are attracted to proteins floating within the surrounding outer cytoplasm. The probability of an activated carcinogen making it across the nuclear boundary, thus, is quite small.

But suppose that it makes it across the boundary. Entrance to the golden city does not necessarily mean gold for the intruder. At this point, the activated carcinogen is able to bind or to attach itself to certain sites on the DNA strand. It then binds to the DNA, attaching itself to the double helix, much like a gangster hanging onto the running board of a 1920s car in a movie about Al Capone. But then, there is another police force in action: the nuclear-immune system. This consists of special molecules in the nucleus that somehow are able to detect extra passengers on the double-helix express. These "repair" enzymes reach out and cut the hangers-on off. In the process, normal DNA units, called nucleotides, also are cut. Other nucleotides then rush in to the surgery site and replace the missing piece. The damage usually affects only one strand of DNA; the other strand, being quantum mechanically complementary to the injured strand, provides a blueprint (or template) for the correct nucleotide sequence to attach to the open wound. The DNA is repaired. The problem here is time: Can the repair operation be accomplished before the cell is scheduled to undergo mitosis into two identical daughter cells (about an hour)? If DNA repair takes place before mitosis, the cancer mechanism is shut down; if not, the daughter nuclei are mutated, each containing a replica of the damaged DNA strand. In this way a mutant gene is created. Each daughter, having survived the split from the other cell, now contains a stable DNA pattern, and all future progeny will inherit the mutation.

But even this "error" in the DNA code does not guarantee cancer. DNA strands in humans make up genes that in turn make up the chromosomes, of which humans possess 46 (except sperm and egg cells, which contain 23 each). In a single cell there are over 50,000 genes, only some of which are capable of becoming oncogenes, or cancer-bearing genes. Most of the carcinogen-producing mutations end up as the normally inactive genes in a cell, or then simply kill or cripple one cell. Only if the carcinogen happens to affect the right gene—that is, mutate the correct site on the DNA strand—will the cell take the first step toward becoming cancerous. The problem is that we don't know which gene is the right one.

Most evidence today indicates that even if that step is taken, another mutation must occur in the now "primed" cell, and the odds against this happening are quite remote. Since only about one cell in a billion becomes primed, the probability of that same cell becoming "secondized" is one in a billion billion. It is here that cancer research enters the game to help explain how such a remote possibility becomes a not-so-remote reality.

Known as the promoter theory of cancer, it states that a second dosage of certain chemicals will react with the primed cells more effectively than with normal cells. Experiments performed by Stuart Yuspa on chemical carcinogenesis related to skin showed that mice did not develop cancer when their skins were painted with known carcinogens. But when these same treated mice were dosed with certain chemicals, tumors were observed in the painted areas. Mice not painted with the carcinogens did not develop any tumors when dosed with the same chemicals. Yuspa calculated the statistics involved, pointing out that when you compare the normal rates of cell growth and cell death, including the chance that mutations will occur at the wrong sites, it begins to become obvious why it takes so long to produce a tumor. His research also indicates that cancers will be most prolific in cells that normally are most prolific themselves, such as skin, gastrointestinal, and uterine-lining cells, and why cancer is hardly ever seen in cells that do not proliferate, such as brain and nerve cells.

What are the implications of these findings for human health? They would seem to suggest that cancer is not an infectious disease and that the conditions for anyone getting cancer must arise from a number of very special effects, each with a low probability of occurrence. Not only must one put carcinogens into one's body, there also must be some special agreement struck between the body's invaded cell and the carcinogen to allow activation to occur and to allow the "local police forces" inside the cell (including the "nuclear police force") to ignore the intruder. These rather unusual "agreements" of getting around the immune system of the cell strike me as a primitive form of consciousness at work. For some reason, dare I say it, the cell *wants* to be invaded! Perhaps cancer is a way of teaching us about cellular consciousness. Perhaps, also, the cure to cancer will arise only through techniques that allow humans to become aware of individual organs and then of the cells inside the body.

It probably seems strange to think of becoming aware of the kidney, lungs, pancreas, liver, and other internal organs; are we aware of any of them now? While most of us can identify with dull aches and sharp pains inside the body, there appears to be no one, with the exception of some Eastern yogis, who is able to normally sense an internal organ.

To bring such a state of consciousness into existence, to become aware of organs and even cells inside organs, may be possible if we know we are attempting to sense. But to become more aware, we need a good model of cellular consciousness. Usually, great developments in science take place when an appropriate model, capable of being tested, is developed. The field of aerodynamics is an example; before people started flying, there were only ideas about flight. In a similar manner, there now exist approximate ideas about cellular consciousness. Once a model is developed (and I suggest that such a model be based on the observer effect of quantum physics), rapid developments will follow. My analogy with aerodynamics is not superfluous. To fly required both the dream that flying like birds was possible and proof through experimentation in the face of many skeptics who believed flying impossible. Bold new visions were attempted.

A bold new approach is just as necessary to reach the new heights of quantum consciousness.

On some level, I believe we are aware of everything going on in our bodies. For instance, if we become ill or have some serious problem with an internal organ, such a malfunction first may occur to us in a dream or other so-called unconscious activity. Some research work with cancer patients, discussed by physician Carl Simonton and psychiatrist Gerald Jampolsky, indicates that life-threatening diseases can be countered by guided imagery. Children suffering from leukemia are taken on guided-imagery programs, where they learn to let go of fear and change their thinking about their illness. Words or phrases such as "impossible," "can't," "should," and "if only," are replaced by "will happen," "can," "will," and "when," respectively. Cancer patients are taught to meditate three times a day, visualizing healthy cells destroying cancer cells.

What is the cure? Can quantum mechanics explain how cancer can be cured? I believe that cancer is not curable in the ordinary sense of the word, like polio or syphilis, for example. Instead, a radical new conceptual basis for the disease must be provided. We will examine such a possible basis in the next chapter.

CHAPTER 29

The Quantum Code of Death

The ideas I wish to discuss here are based on some early work by physicist Per-Olov Löwdin. As early as 1964, he speculated that the reason we all die of natural causes is quantum physical in origin. He went on to say that death, problems in mutation, aging, even tumor growth, ultimately are caused by a quantum physical process in DNA that results in genetic miscoding. This error is produced by proton tunneling in the hydrogen bonds found between the base pairs that comprise the steps of the DNA spiral. In this chapter I hope to make Löwdin's thoughts clear to the reader, and then offer my own ideas as to how consciousness can alter the genetic scrambling caused by quantum mechanical proton tunneling.

Proton tunneling is a quantum physical process having no intuitive counterpart in our everyday world. Yet tunneling is a common occurrence for any atomic or subatomic particle, not just for protons. The particle can be visualized as a tiny marble rattling in a glass goblet. Suppose the goblet rests on the table. According to quantum physics, if the goblet and the marble were atom-sized, the marble would begin to quantumly rattle inside the goblet in an attempt to free itself from the glass boundary surrounding it. Not having enough energy to climb the glass wall, it would simply rattle back and forth inside. However, in quantum physics atom-sized marbles do not remain solid little marbles all the time. As they rattle, they change into waves, and as waves they are allowed to do something they cannot do as particles: They can "shine through" the glass wall that surrounds them, much as light waves reflecting the thumb that holds the full water goblet pass through the glass to our eyes. Once the quantum wave has passed through the glass, as soon as it is observed it returns to its guise as a particle. This kind of tunneling occurs in every microchip of every computer circuit manufactured today.

It also may be the chief cause of human suffering and diseases—particularly the diseases I label as quantum mechanical.

236

In the human cell, proton tunneling takes place in the tiny molecules controlling our inherited characteristics—our DNA. Again, each DNA molecule consists of two long molecular strands, composed of sugar pentagonal-ring molecules interspersed with phosphate connecting-link molecules, spiraling about each other, as stated earlier, like the banisters of a spiral staircase (see Figures 24 through 27). Connecting these strands are molecular steps. Each step contains a pair of molecules taken from a set of only five base molecules, called, for simplicity's sake, A, C, G, T, and U. These molecular bases will not couple or pair with each other indiscriminately: A normally will pair only with T or U, but never with G or C. Similarly, G will pair only with C.

The reason for this selective pairing has to do with the nature of proton chemical bonds themselves (see Figure 28, page 240). They exist as "corridors," connecting one molecule with another, with each corridor containing a single proton—an atom of hydrogen without its electron.

FIGURE 24.

SUGAR PENTAGONS OF LIFE

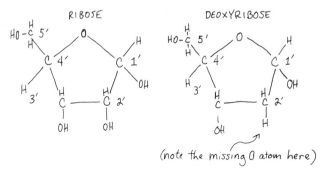

(note the missing O atom here)

FIGURE 25.

PHOSPHORIC ACID:
THE ACID QUEEN OF DNA AND RNA

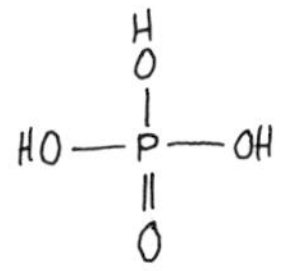

FIGURE 26.

THE BACKBONE OF QUANTUM MOLECULAR LIFE

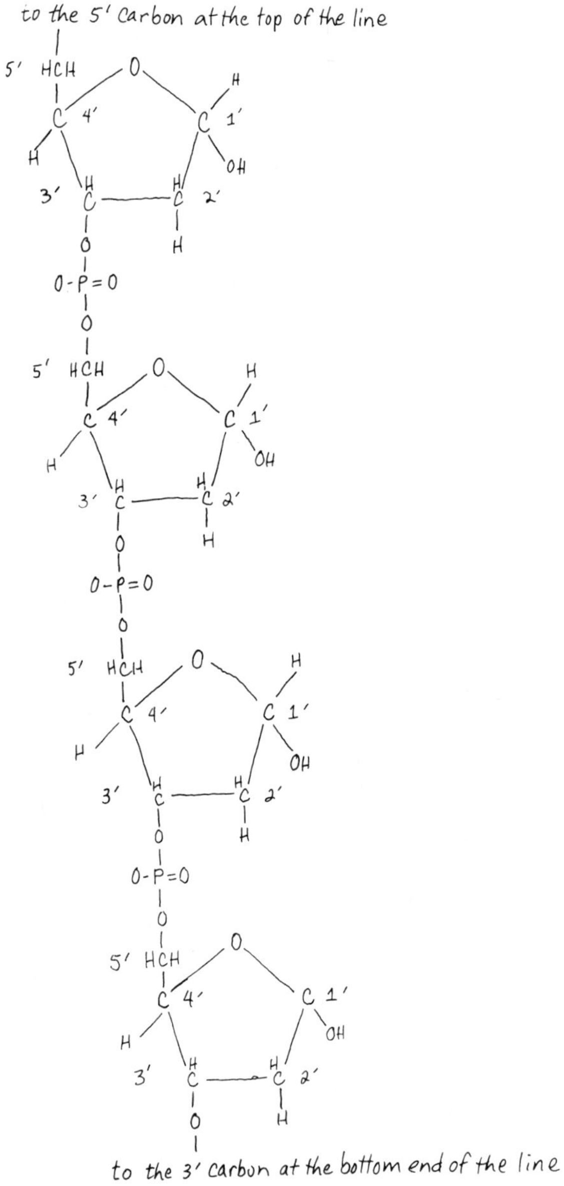

to the 5' Carbon at the top of the line

to the 3' carbon at the bottom end of the line

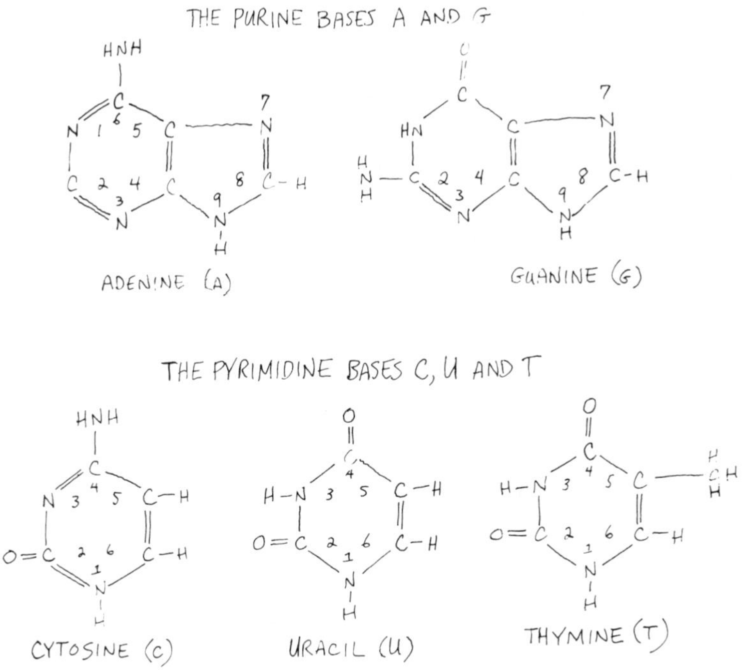

THE PURINE BASES A AND G

ADENINE (A)

GUANINE (G)

THE PYRIMIDINE BASES C, U AND T

CYTOSINE (C)

URACIL (U)

THYMINE (T)

FIGURE 27.

Coming from an A molecule are two corridors—called the top and middle corridors—the top containing a proton, the middle corridor not. Similarly, a T base also has a top corridor and a middle, only this time the middle contains the proton. The same applies to the U molecule. Thus, when A and U (or T) come together, there is a tendency for their respective corridors to bond, trapping a single proton between them in each corridor, where, like atom-sized marbles in glass goblets, they rattle back and forth. Between A and T, both protons rattle in their respective corridors.

In the C-G bond, there are three corridors: The proton in the upper corridor is associated with the C base, while those in the middle and lower corridors are associated with the G base. Again, the C-G base connects up respective corridors, allowing all three protons to rattle in them like mysterious ghosts in haunted hallways. Such is the nature of the proton chemical bond.

THE C≡G TRIPLE HYDROGEN BOND (fig. a)

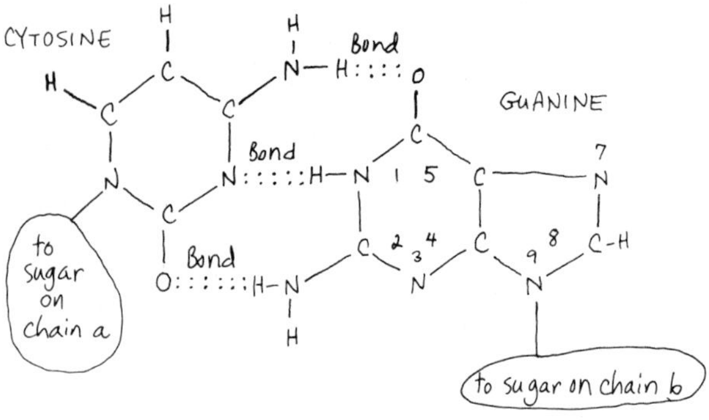

THE T=A DOUBLE HYDROGEN BOND (fig. b)

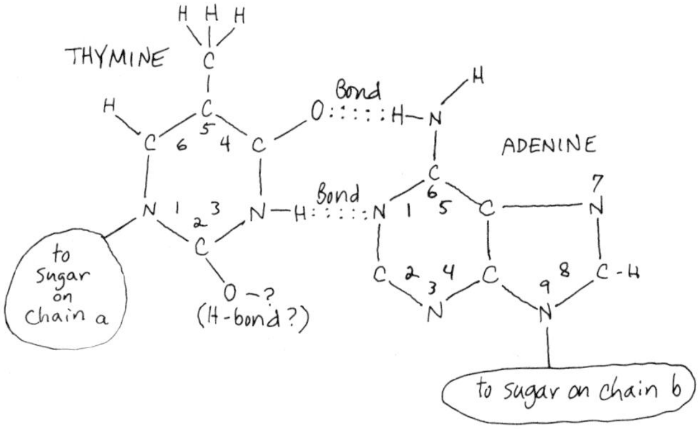

THE U=A DOUBLE HYDROGEN BOND (fig. c)

FIGURE 28.

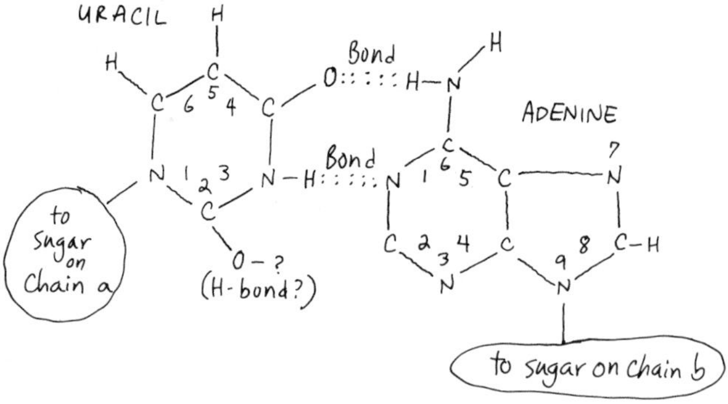

The Breakup of the Quantum Mechanical DNA Computer

The tie between base pairs A and T arose as a two-corridor or double-hydrogen bond, and the tie between base pairs C and G arose as a three-corridor or triple-hydrogen bond. In each bond a single proton is closely attached to one of the bases but also is attracted to the other. Thus, it rattles back and forth. To simplify the concepts involved, these base-pair connections are presented in Figure 29 below. The bottom row of the figure gives a rotated image of the top row when each base is turned around, back to front, along the vertical axis in each graph.

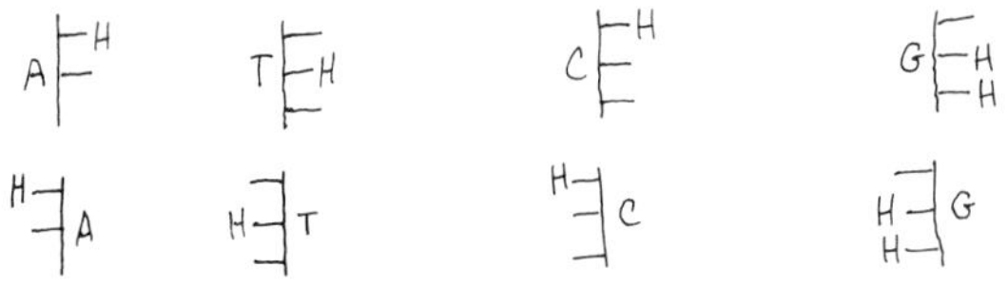

A simple view of proton bonds in bases

FIGURE 29.

The base A has two bond possibilities, with one proton (H) attached to the upper bond. The base T has three bond possibilities, with a proton attached to the middle bond. Base C, likewise, has three bond possibilities, but its proton is attached to the upper bond. Base G, also with three bond possibilities, has a proton attached to both the middle and the lower bonds. With Figure 30, it is easy to see why A and T fit together, and also why A does not fit either C or G.

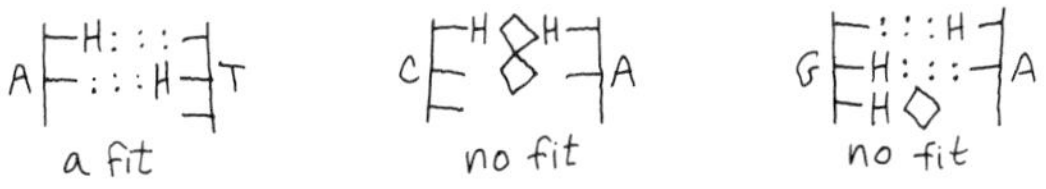

a fit   no fit   no fit

How A-T, C-A, and G-A bases fit or don't fit together

FIGURE 30.

A and T fit because A supplies the proton to the upper bond, while T supplies the proton to the lower. It's like having a double-key security system at the ends of the corridors. Each base supplies a key (H) and a lock (::::) or a tendency to hold on to a proton. C and A do not fit because while each has a key (H) in the upper bond position, they do not have locks (:::) to form the bonds—the keys are in the locks. In effect, the protons repel each other. A and G do not fit because of the presence of H in the lower-bond position of G, which repels A. If that same H were not present, G and A would fit together, thus causing a mutation, or a quantum mechanical error (Figure 31).

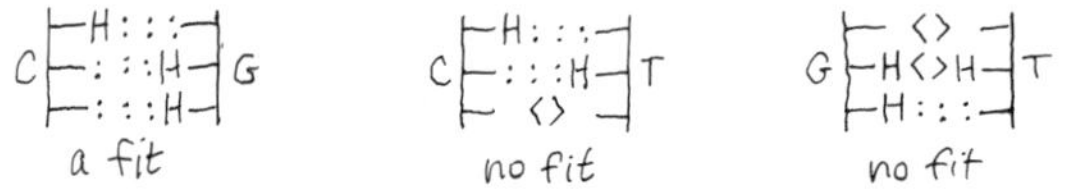

How C-G, C-T, and G-T bases fit or don't fit together

FIGURE 31.

C and G fit together in a similar manner; this time, G supplies two keys and one lock to C's one key and two locks. C and T and G and T mismatch (<>) because either there is bond repulsion caused by the lack of a proton (-<>-) or proton repulsion (-H<>H-) caused by two protons in the bond. Therein lies the reason for the genetic code's ability to reproduce itself because of the complementary fit.

The Hydrogen Bond Revisited

A word about the hydrogen bond itself. I have drawn it as -H:::- or as -:::H-. The symbol ::: stands for an attraction for the proton coming from the empty position. It is possible, therefore, for -H:::- to spontaneously tunnel into -:::H-. In this case, the proton has tunneled from one attractive site on (or side of) the bond to the other. This could not occur in classical physics because between the two sites there exists a potential energy barrier. We can draw this as in Figure 32.

In the figure we see both the upper and the lower barriers existing between bases A and T, comprising the double-H bonds between these two. Also shown are the three potential barriers existing as the triple bond between C and G. In each case there is a fit corresponding to a single proton trapped in one side or the other of the hill. The proton should not be able to make it across the hill unless it has enough energy to climb it, that being the case only for a classical particle, which, of course, the proton is not. Depending on the height of the barrier and the bond length separating A from T, the proton can tunnel through the hill. Take the proton out or put one on each side of the barrier and the bond becomes totally repulsive and does not form.

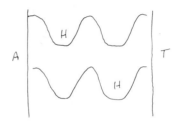

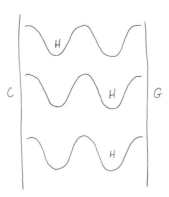

FIGURE 32

Protons (H) in quantum valleys in the A-T and C-G base pairs

According to quantum physics, a proton can and does tunnel from one valley through the hill and arrive at the other side. In fact, according to the uncertainty principle, it has a small but finite probability of existing there already!

How Genes Miscode, Producing Genetic Error and Death

If a proton manages to hop from one bond to another, or to form with its proton in the wrong bond in a particular base, it is said to change form and become a *tautomer* of the original base. A few possibilities for tautomeric forms of the bases A, C, G, and T are shown in Figure 33.

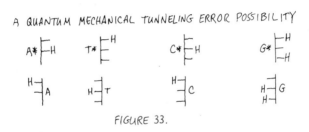

FIGURE 33.

A comparison with the normal base forms can be drawn by looking at the top and bottom lines of the figure. For example, instead of the proton residing in the top bond of A, it resides in the middle bond of A*. These tautomeric forms, consequently, do not combine with the same complementary bases as the normal forms do. (The reason is shown in the next set of figures.) For example, if an A* forms from an A-T split, during DNA replication, the A* will no longer bond to another T; it will bond instead with a C. Further splits then will introduce base-pair errors (Figures 34 through 37).

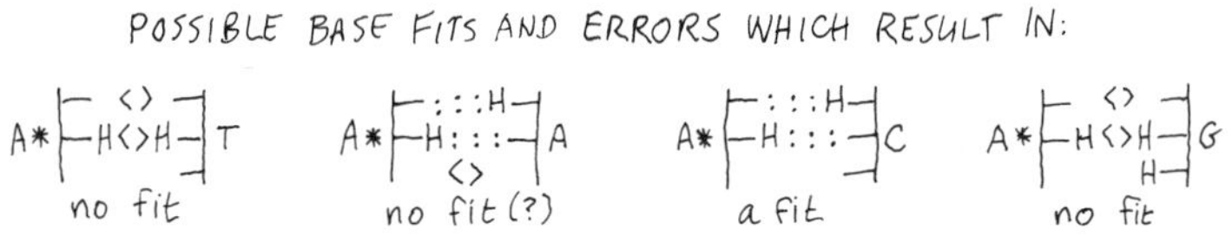

POSSIBLE BASE FITS AND ERRORS WHICH RESULT IN:

FIGURE 34.

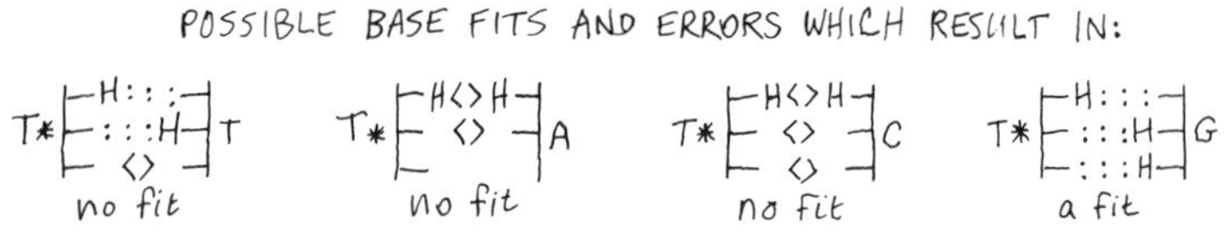

POSSIBLE BASE FITS AND ERRORS WHICH RESULT IN:

FIGURE 35.

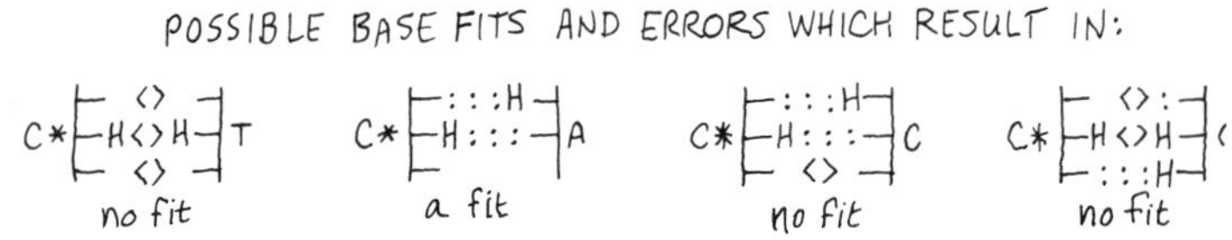

POSSIBLE BASE FITS AND ERRORS WHICH RESULT IN:

FIGURE 36.

POSSIBLE BASE FITS AND ERRORS WHICH RESULT IN:

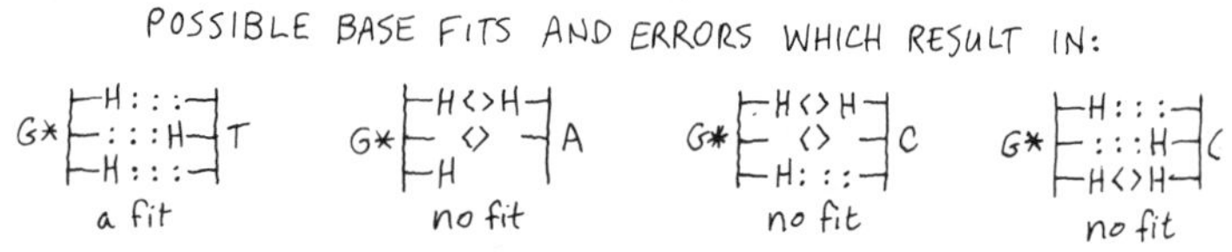

FIGURE 37.

A normal growth tree from A-T splittings from a single DNA strand is shown in the next set of figures. With the first split of the A-T bond (see Figure 38), each partner picks up a complementary partner, thus further engendering new A-T base pairs. However, from an A*-T split, as we see in Figure 39, both A-T pairs and C-G pairs occur after two generations.

ERROR TREES
(FROM LITTLE PROTON TUNNELS DEATH AND CANCER GROWS)

A NORMAL DNA SPLIT

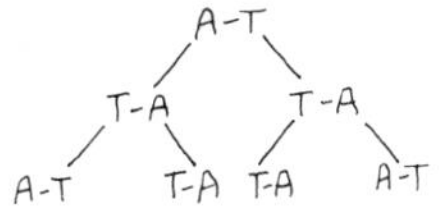

FIGURE 38.

AN ABNORMAL GROWTH FROM AN A*-T

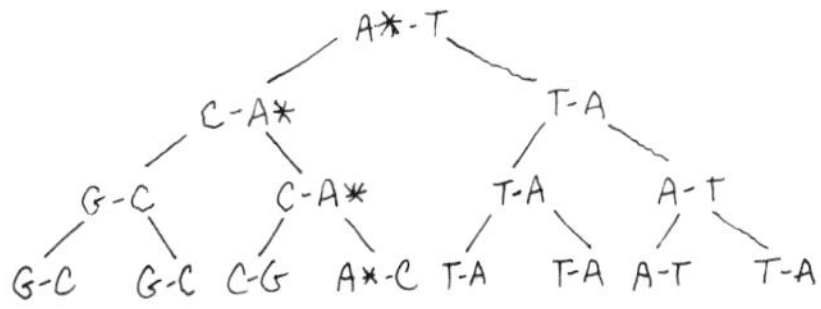

FIGURE 39.

Here another possibility arises. If a tunneling occurs in both the upper and lower bonds of an A-T pair, a new pair A*-T* is produced. When this pair splits (see Figure 40), there is no hope of ever regaining an A-T bond again. The A* will bond only with a C and the T* only with a G, resulting in C-G pairs.

AN ABNORMAL GROWTH FROM AN A*-T*

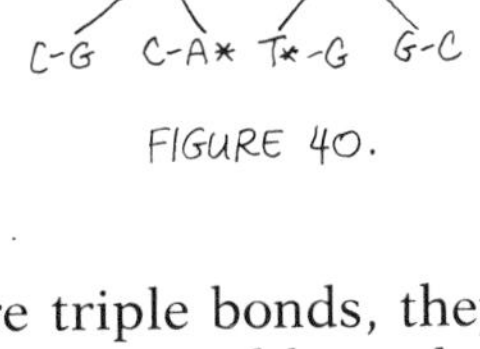

FIGURE 40.

Since C-G bonds are triple bonds, they are stronger than A-T bonds, and, therefore, more stable and more resistant to spontaneous splits. If cancer cells have C-G pairs replacing A-T in the right places of their DNA strands, it is no wonder that they are able to resist eradication and multiply so prodigiously. It is possibly the conversion of A-T pairs into A*-T* that ultimately results in the aging processes that steadily sap the vitality of human organs.

A similar line of reasoning holds true for C-G pairs. If a C becomes a C* through proton tunneling or some other mechanism, then the C*-G pair will not produce C-G pair-bond offspring only. C* will bond only with A; thus, C*-G will produce A-T offspring as well as C-G pairs. There is a probability that C-G also will become C*-G* by having two protons tunnel simultaneously through their respective bonds. This results in A-T bonds that are double bonds and that thereby are weaker in counteracting perturbative forces acting upon them. These bonds, located in the wrong places in the daughter DNA strands, could cause the cellular degeneration and death of the cell.

A Quantum Look at Disease

The concept of *quantum physical* as opposed to *classical physical* disease constitutes a new look at health. There is no single material cause of cancer, atherosclerosis, osteoarthritis, diabetes, emphysema, or cirrhosis, and in my view, looking for cures, in the conventional sense, is foredoomed. Although no single material cause exists for any one of these diseases, a single, nonmaterial cause does. This cause may be consciousness acting through the quantum physical observer effect. Consequently, although much can be done in the way of disease prevention by proper diet and exercise, just postponing or warding off a disease does not constitute a cure.

To find cures for these quantum mechanical diseases, one must use quantum physics as a basis. This means that such events as proton tunneling should be suspected as the causative agents in these diseases. If this turns out to be so and if it also is true that human consciousness can alter physical matter simply by observing it, then techniques may be developed better able to cure these maladies.

In the next chapter, I explore some possibilities for the body quantum—the observer—in action.

CHAPTER 30

The Mind as Quantum Slayer, the Mind as Quantum Healer

With a knowledge of quantum physics, it is possible to construct a reasonable model for the types of diseases that result in a loss of vitality. The human cell in this model represents a compromise; it is a picture of quantum physical complementarity, a battleground between forces of order and chaos.

The electrical forces that tend to align and to correlate movements of electrons and protons that make up the hydrogen or proton bonds between base pairs in DNA are forces of order. Through quantum physics we now know that whenever two systems that were previously uncorrelated interact, they form a common quantum wave function, or a qwiff. The information contained in a correlated system is always greater than the information contained in a system that is uncorrelated. Correlations, therefore, tend to lock in information and to minimize entropy in the respective systems.

The Mind: Breaking Correlational Bonds

When observation occurs, it tends to destroy correlations, taking information from the bonds of correlation. This radically alters the qwiff, while momentarily leaving the physical situation the same. At this juncture, probability patterns are altered radically, and with a change in pattern, there is a greater likelihood for mutation or a sudden transformation in the DNA code. With greater body awareness, it follows, there exists a higher probability for induced change and transformation. And therein lies the cure and the curare.

Because the mind may enter into the DNA bond, it may alter its qwiff pattern, resulting in a greater possibility for proton tunneling. This tunneling can lead to illness. It may also enter whenever proton tunneling has occurred spontaneously, through luck alone, altering the probability so that the proton is brought back where it belongs. This retains health. By becoming more mindful, we may either increase our vitality, or sabotage it.

Protons at the Whims of War in the Love/Hate Dance of the Hydrogen Bond

Since it is the hydrogen bond, -H:::- or -:::H-, that interests us most in the nature of quantum mechanical diseases, it is useful to consider in greater detail what it is and how it forms. A hydrogen bond is made from a pair of electrons and a single proton. If we look at the top bond of an A-T pair, it looks like -H:::-O. The single − represents a pair of electrons holding the proton (H) to it. The :::- represents both the potential energy "hill," as well as the attractive force on the other side of that hill. In Figures 41 through 43, we see the pair of bonds in the A-T base pair. The new detail is that the bonding energy on each side of the barrier is not the same. Consequently, the proton tends to be more on one side of the hill in the lower valley than on the other side in the upper valley.

In the normal arrangement of the A-T pair, the proton is located in the lowest valley, indicating a relatively stable position.

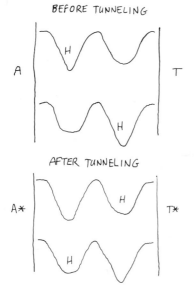

FIGURE 41.
How an A-T pair tunnels to an A*-T * pair.

BEFORE TUNNELING

AFTER TUNNELING

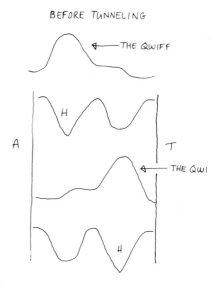

FIGURE 42.
Quantum waves (qwiffs)
before tunneling in A-T base pair

BEFORE TUNNELING

THE QWIFF

THE QWI

248

FIGURE 43.

Quantum waves (qwiffs) after tunneling in the A*-T* pair

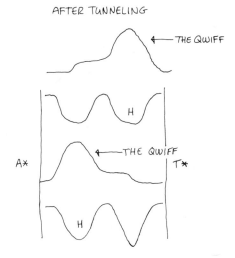

AFTER TUNNELING

THE QWIFF

H

THE QWIFF

A*   T*

H

After a simultaneous tunneling of both protons in the upper and lower bonds, the protons are in the higher valleys of the A*-T* base pair. The existence of the high-valley sides constitutes a threat to the life of the DNA's ability to reproduce itself without mishap. The greater the difference between the heights of the valley floors in each bond, the smaller the probability that tunneling will occur. However, as time goes on, the qwiff, existing across the bond, will change, and so too the probabilities for tunneling.

This phenomenon is quite similar to musical resonance. When a singer in the familiar Memorex commercial hits a high C, a glass on the other side of the room suddenly cracks and explodes. This is caused by resonance, wherein the vibrations of the singer's vocal chords just happen to hit a resonance frequency of the glass. The glass then begins to vibrate, finally breaking as the tone coming from the singer to those vibrations already present in the glass.

In a similar manner, the proton vibrates in the lower-valley side of the bond. Since there is also a resonance frequency for this vibration to occur in the upper-valley side of the bond, eventually the qwiff builds up in this region. As it does, the probability for finding the proton on the low-valley side decreases as the probability for finding it on the high-valley side increases. When the odds reach 50–50, the likelihood of the sudden appearance of the proton in the high-valley side is greatly enhanced. Eventually, it just happens.

The frequency for this occurrence depends highly on rather complex environmental factors. It probably depends on temperature. It also depends on the height of the barrier or hill and on the depths of the valleys. So too it depends on the neighboring bonds, their net charges—indeed, the entire electrical environment of the bond itself. The main problem is whether tunneling time is long or short in comparison with replication time, and that may depend on your own mind.

Where Was Your Mind When the Proton Tunneled Through?

To grasp how our minds enter into the bonds of war and peace that exist between the strands of our DNA molecules, we need to examine the basis for all quantum physical processes—the quantum physical probability amplitude for anything to happen. This amplitude can be thought of as the "loudness" of a quantum wave form in much the same manner as the amplitude of a sound wave corresponds to its loudness. The larger the amplitude, the louder the sound. In quantum physics, the larger the amplitude, the higher the probability that something actually will take place. The symbol for this amplitude is written $<a \mid b>$. This peculiar symbol stands for the probability amplitude that if some event, b, is known to occur, $\mid b>$, and the possibility, a, may occur, $<a \mid$, then the possibility that a occurs, given that b has occurred, is $<a \mid b>$. Usually this amplitude changes with time, meaning that sometimes it is higher than at other times. If a physical situation is vibratory or oscillatory in nature, there will be a frequency associated with this amplitude. The higher the amplitude, the greater the frequency.

If a and b represent the locations of the protein in the high and low valleys, respectively, then the tunneling frequency depends strongly on the quantum probability amplitude, or quamp $<a \mid b>$. This quamp is the quantum amplitude probability that a proton initially with position b is later found to occupy position a. To determine this, one looks at the quantum waves (or qwiffs) corresponding to before and after tunneling. In effect, one multiplies them together and adds up the products. A look at the qwiffs for the before- and after-tunneling positions in the upper bond of the A-T pair is shown in Figures 42 and 43. To see how this multiplication is calculated, consider Figure 44.

To compute $<a \mid b>$, multiply the magnitude of the upper qwiff at one point in space by the corresponding magnitude of the lower at the same spatial point, then progress to the next point and repeat. Finally, one adds up all the products. Of course, this is

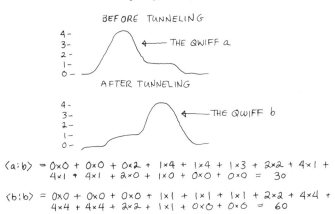

FIGURE 44.

Quantum waves (qwiffs) and their probabilities

BEFORE TUNNELING

THE QWIFF a

AFTER TUNNELING

THE QWIFF b

$\langle a:b \rangle = 0 \times 0 + 0 \times 0 + 0 \times 2 + 1 \times 4 + 1 \times 4 + 1 \times 3 + 2 \times 2 + 4 \times 1 + 4 \times 1 + 4 \times 1 + 2 \times 0 + 1 \times 0 + 0 \times 0 + 0 \times 0 = 30$

$\langle b:b \rangle = 0 \times 0 + 0 \times 0 + 0 \times 0 + 1 \times 1 + 1 \times 1 + 1 \times 1 + 2 \times 2 + 4 \times 4 + 4 \times 4 + 4 \times 4 + 2 \times 2 + 1 \times 1 + 0 \times 0 + 0 \times 0 = 60$

only an approximation, but it gets the point across. One can compute <b | b> in a similar manner. Here, one multiplies the qwiff, b, by itself, term by term, then adds them up. The total for <b | b> is 60, while for <a | b> it's 30. Thus, <b | b> is always greater than <a | b>. Looking at these products, you see that in the case of <a | b> there are 7 zero-products in the sum because where qwiff b is large, qwiff a is small, and vice versa. There are only 5 zero-products in <b | b>. Thus, it's where the waves overlap that produces any value at all, and this depends on how big each qwiff is in the forbidden region of the central hill (see again Figures 42 and 43).

The greater the overlap, the higher the probability for inducing a tunneling transition. The question here is, What do we do to ourselves to enhance <a | b>? In other words, what are the factors that lead to relatively larger probabilities of forbidden-zone penetration?

To grasp the answers, we need to remember just what these amplitudes stand for. They are probability amplitudes, after all, not physical facts, and probabilities refer to chances, not certainties. Whenever <a | b> is large, the chance that a proton tunnels from position b to position a is large also. But that does not mean that the event will occur, for no event occurs without the body quantum effect. Someone must observe an event in order for that event to really take place. If no observation takes place, the probability amplitude simply will continue to oscillate. The proton will exist in wave form, with amplitudes on both sides of the hill. The

251

bond remains in place. If observation does occur, and the proton is "spotted" with certainty on one side of the barrier or the other, the bond snaps like a broken rubber band.

Here is where "you" enter. By "you," of course, I mean "that which observes" inside of us all. This "you" constitutes a form of intelligence—an all-seeing eye. This "eye-I" is responsible for the sudden change in a probability amplitude that constitutes the changing of probability into certainty. Just as a baseball batter, noticing that a pitcher tends to pitch high balls more than low ones, alters his batting stance to swing higher, "you" can alter your frequency of noticing when the probability of proton tunneling is higher. By weighting your swing of consciousness in this manner, you can pay attention to high-probability-event possibilities, and more or less create the tunneling actuality. On the other hand, if you don't look at all or observe only when the probability amplitude is low, the bond will not break as often.

Growing decrepit and becoming less healthy may be caused ultimately by our spirits, our "eye-I's," observing more and thus causing more damage to our genetic structures. Illness and death may be nature's way of becoming more intelligent. This is a cosmic joke, yet it is deadly serious.

PART EIGHT

Transforming

We have all seen miracles. People we know suddenly change their bodies, their outlooks, their whole personalities. Athletes reach new goals of excellence; people who, heretofore, have exhibited no talent for anything suddenly transform into talented performers. One way to grasp such divine transformations is found in the parallel-world concept of quantum physics.

A parallel world in quantum physics is an alternative explanation of the observer effect. Instead of a sudden alteration of a quantum probability into an actuality, the parallel-world hypothesis posits that each possibility actually occurs in a separate world. Each world is made up of a continuous stream of events; each event may be a branch point, connecting one world to another. Since there are a countless number of possibilities for each event, the number of worlds is countless. In some of these worlds, the worlds of which we are not directly aware in this world, there exist duplicate events making up a parallel "you." This parallel "you" is aware of you in much the same manner as you are aware of "you." Should the "you" in a parallel world appear to the you in this world as a dream being, then the this-world you would also appear as a dream being to the "you" in the parallel world. It is through our dreams and alternate states of awareness that the you and the "you" are able to communicate.

The parallel or dream world surrounds every one of us as

an aura, a dream body. This dream body is as much a reality to the real or physical body as the qwiff is to the physical or material substance of everything in the universe.

The idea here, too, is that our normal body awareness is actually composed of an infinite number of separate awarenesses—each one aware of some particular aspect of the body. Those aspects that we take as part of our normal awareness, such as the awareness of body heat or cold or of sensations in our hands and legs, compose what we call "waking consciousness." We call this awareness our universe or our world. However, unconscious to all of this are the other parallel worlds. In our present state of culture and technology, we probably can become aware of these only through dreams, intuitions, and flashes of such psychic events as déjà vu.

The so-called miracles mentioned above probably take place when there is a great deal of agreement among our separate selves—when we cease fighting our selves and behave consistently with our dreams. Such agreement can occur only when there is a "paying of attention" to messages coming in from our other selves in these worlds. Each "you" in the parallel world then would have the same sense of what is to be accomplished. This entails, practically speaking, becoming aware of the fringe areas of our personalities, our dark sides as well as our light. And this takes us into the realm of the mind.

In this view, illness takes on a different nature: It is a message from the dream body to the physical body that indicates that transformation is taking place. Consciousness is the bridge that enables you to tune in to the dream body and determine what choices are available to you. The means for this awareness lies with the proprioceptors, the sensory inputs that feed back information from the muscles to themselves.

The interaction between you and a parallel-world "you" can be thought of as the ego structure of Freudian psychology. Probably the most important conceptual basis for health is the transformation of the human ego. The ego is explained as a Freudian concept arising from the id. In my quantum model, the id is a repository of fundamental quantum physical processes. These processes involve both the death wish and the life forces. These forces arise from the processes of annihilation and creation of matter in quantum physics. The ego is explained as a process of self-contraction by several spiritual teachers. I will use this concept to make a simple quantum physical model of the ego. Pain and pleasure, and stress and health as well, are seen as a function of the actions of the observer on the boundary regions of the physical units comprising the body. Ego expansion is a release of tension, through thought, that produces pleasure but that is not always easily accomplished.

The whole mind-body—a model based on the quantum physics of the ego—provides a model of health both for the individual and for the whole planet.

CHAPTER 31

Messages from a Parallel Universe

Arnold Mindell, a key figure in the revolutionary field of dream and body work, is a psychotherapist and Jungian analyst. His background also includes a master's degree in physics and a Ph.D. in psychology. In his book *Dreambody*, he explains that the human body is not only a real physical body, as envisioned by medical practitioners and classical physicists, but also a dream body. This dream body is a receptacle of dreams, visions, and myths that we carry with us from childhood. It receives messages from the unconscious, which often appear to us as unwilled bodily movements, pains, and illnesses, some even resulting in our own deaths.

How Your Real Body Is Also a Dream Body

The idea of a dream body originally came from the work of Carl Jung. It has been touched upon by many ancient religious systems, and it is employed today by neo-Reichian and gestalt therapists, as well as body therapists. Mindell points out the Jungian concept of the dream body has radically transformed his practice of psychology. The dream-body concept arose from Mindell's work on the psyche-matter interaction, a study that also was quite attractive for both Jung and Freud. When Mindell became mildly ill himself, he wondered why the theoretical developments of psychology that he was using were of no use in curing himself of his illness. Later, after many thousands of hours of dream analysis and much labor over imagination techniques and association, he turned to the body and its phenomena.

In Western medicine, the body is considered as separate from the mind; it is seen as a network of physical processes, much like a machine composed of separate parts. Its behavior, it is believed, can be determined by understanding the nature of its individual parts.

The body is also viewed this way in classical physics. The concept of understanding the physical body is the ultimate reduction of the body's workings to particles of matter, moving and interacting in space and time. The notion of a real body, thus, has implied within it the notions of classical, deterministic physics.

Quantum physics indicates that the role of the observer plays a major function in the reality of those processes through the principles of indeterminism and complementarity. The body is not composed of separate parts, each behaving in a determined manner; instead, it is made up of complementary processes that make tradeoffs. Observation enters and decides what can be observed, at the expense of complementary observations. A change in the way a person observes his or her body alters the energy of the body processes themselves.

The concept of the dream body goes beyond the real body in the same way that quantum physics goes beyond classical physics. The chief difference between the two branches of physics lies in the conceptual framework against which each pictures the real world. Classical physics holds that there is a real world out there, acting independently of human consciousness. Consciousness, in this view, is to be constructed from real objects, such as neurons and molecules. It is a byproduct of the material causes which produce the many physical effects observed.

Quantum physics indicates that this theory cannot be true— the effects of observation "couple" or enter into the real world whether we want them to or not. The choices made by an observer alter, in an unpredictable manner, the real physical events. Consciousness is deeply and inextricably involved in this picture, not a byproduct of materiality. At this point quantum physics shows that there are two major ways of dealing with the manner in which consciousness is involved in physical processes.

With the first, consciousness exists outside all physical events; it enters whenever an observer makes a choice. It then acts by "forcing" the quantum system to enter into one or another possible real situation. The outcome of this "forcing" is the appearance of something physical, such as the location of a particle in a measuring instrument.

With the second, known as the parallel-world theory, consciousness does not do this; instead, the observer's consciousness "splits" when an observation takes place. That split is unnoticed by the observer because the observer also splits: One observer sees and records the particle at one location; the other sees and records the particle at another location. Each observer remains unknown to the other until yet another kind of observation occurs. (Albert.) Each observer exists in a parallel world; each feels that his or her consciousness "forced" the particle to enter into a measuring instrument that records its position. And each observer is unknown to the other. However, here another kind of measurement can take place. It involves both worlds, and it can be recorded only if both worlds are taken to be a single world. This measurement is complementary to the separate world measurements. In this way, a person in one world can become aware of his or her alter ego in the other world.

In a similar manner, the dream body can be best conceptualized in terms of the parallel-world picture of quantum physics. The dream body enters this physical body when something is slightly off. This often takes place when illness jars our normal consciousness.

The usual question of the body when it is ill is, What is wrong with the body? The idea that something is wrong implies that there is a right answer. Thus, adjusting the body chemistry through drugs or therapy attempts to alter the physical circumstances of the "diseased" processes and bring them under control or in line once again.

No one really knows just how a drug cures the body of an illness. We know, of course, that an antibiotic will alter the chemistry of a bacterium, eventually destroying its ability to reproduce. But what causes the body to resist an infection in the first place? To say that we all have immune systems really doesn't explain it, and instead probably relieves us of any need for explanation: We take the medicine, we trust the doctor. Then the miracle we call "nature takes her course" happens, and in a few days we are cured.

We picture our real physical bodies as ill or diseased. The very notion that there is only *a* real body probably contains within it the very seed of disease.

But in a parallel-world vision of the body, illness can be thought of as a communication between the body and the dream body, really, communication between one parallel world and another that results in a conflict—a disruption of some pattern in our lives. This is not the only way to receive a message from a parallel world, but for most of us, in our present states of awareness, it may be the closest we can come to a form of awareness beyond our usual concerns.

A further clarification is needed here. In a classical-physical world view, the body responds to programs much in the same way that a computer responds to programs. Feed in a certain set of data and a means for manipulating that data, and the body will do such and such. In the quantum-parallel-world view, the same holds true for all bodies in the parallel worlds—that data input and its programming affect every parallel-world body. But no one body can predict, except statistically, just what will happen in any one of the parallel bodies. Some will be healed quite quickly; some will become sicker. The chances are that your body will be a member of the set of bodies in which no instant miracle occurs. The other parallel bodies then will communicate to you that they are "out of phase" with you. The sicker bodies will say that they are getting sicker, the healing ones will say that they are getting well, miraculously. You will be somewhere in between, listening to their messages.

In this sense you won't be able to predict exactly what your body will do, but you may be able to alter the progression of your disease by learning to tune in to the healing alter-world bodies. This probably happens through an emotional release—a letting go of factors and conditions in your life that produced the stress that usually surrounds disease. The healed bodies in your parallel worlds were healed because they had no vested interest in remaining ill. You may have such an interest; the battleground for such interests is the unconscious mind. Their various messages are received. With effort you can tune in to positive healing and loving messages, or you can tune in to disasters. To some extent, depending on the state of your "ship" (just slightly damaged, sinking, or beyond repair), you can alter the progression of your illness and probably heal yourself.

My point is that the body is not real in the same sense that a particle of physics, such as an electron, is not real. The body does not possess well-defined boundaries and cannot be predicted to move or respond in specified ways. The body, instead, is perhaps viewed best as a confluence of agreement brought forward by a resonance of parallel processes, each process taking place in a parallel world. What we call reality, in effect, is made up of the infinities of processes, each taking place in separate and noninteracting worlds.

One interpretation of the body quantum is that the body is composed of vibrations and fields. Just as an electric field is imagined to permeate space, the dream body permeates the same space as the real body. And just as the electric field cannot be seen in and of itself—only its effects on matter can be observed—so too the dream body is witnessed in its effects on the real body.

This idea of the dream body as field intensity also can be thought of as a double, or *doppelgänger*. In an episode of Rod Serling's *Twilight Zone* called "Mirror Image," a highly rational woman, "not very imaginative, not given to undue anxiety or fears, or, for that matter, even the most temporal flights of fancy—a woman with a head on her shoulders," comes to a bus depot and becomes highly disturbed when events beyond her control but in which she is intimately involved occur. She suspects that the bus station is run by lunatics when the ticket taker snappishly tells her that her suitcase has been checked already and that she has repeatedly asked when the bus will arrive. In the washroom, the attendant tells her that she has been there just a moment before. Yet she has done none of those things.

A man, also waiting for the bus, offers a sympathetic ear. When the bus arrives, the two of them begin to board. Suddenly she shrieks in horror and flees back into the station after seeing *herself* already on the bus. The man misses his bus and listens as she explains what has happened: She has a double; a mirror image of herself from a parallel world somehow has slipped into this world and must take her place in order to survive.

Certain that she has gone bonkers, the man summons the police, who come and take her away. A few minutes later he regrets his decision: Chasing an elusive figure down the street who he assumes has stolen his suitcase, he is horrified when the figure turns around as it runs away and grins mockingly at him. The grinning face is his own!

Serling's stories are still popular today, perhaps even more so than when they first appeared over 25 years ago. The idea that we have another body, a dream or spirit body that mirrors our own and is fighting with us for its survival, shocks us but, at the same time, rings true. In ancient India, this double is called *linga share-irah*; French spiritualism calls this phenomenon the subtle body, *érisprit*. Shamans, mediums, and healers see the human body as surrounded by an aura—the presence of the dream body. The white glow surrounding Christ in all his pictures is the dream body, as intuited by artists.

What Your Illnesses Really Tell You

In the winter of 1978, I was invited to give a series of lectures to the Philosophical Research Society in Los Angeles. My topic was "The Conscious Atom," which promised to bring many students of mysticism and metaphysics to the gathering. And indeed it did. One was a man who impressed me with his experiences dealing with his double. I will call him Drew.

Drew was one of those students who really had more to teach than to learn. He recently had had a severe bout of illness, and even as he told me about his double, his health still was not good. He told me the following:

Imagine that you are on board a ship heading somewhere. Although you are at the bow and are able to walk from stem to stern and see where the ship is heading, you are powerless to change its course. You fear that something is wrong, but no one is on deck, and no one, it seems, is at the helm, either. No one is around for you to call or appeal to.

As you make your way around the ship, you realize that you are locked out. You cannot find any open doors or even portholes. The ship is grinding ever onward to some destination and you appear helpless. But then you hear something. You put your ear near the boiler-plated door leading inside the ship and you hear a voice inside. The voice is crying for help, for guidance.

As you approach the door, your senses are assaulted by a stench coming from the other side. The voice, however, is quite strong and claims to be steering the ship. But the voice coming from the darkness cries that he cannot see where the ship is heading. He is asking for someone to open the door, to let him see where his actions are taking the ship.

You scream at him through the door, "I am here outside. I can see where the ship is going. I can help you steer the ship, but I am helpless to control it myself. Can you hear me?"

He cries back, "Is there anyone out there? I have been in the dark of the hold so long now I am not even sure there is an out there out there. Is anyone there?"

You yell back at him, "I am here, listen to me." But, alas, he doesn't hear you. You know why he cannot. The stench inside tells you that the hold is rotting. The noise of running the ship and the inattentiveness of the person inside to the outside world he cries for keep him from hearing. You cannot go into the hold even if you could manage to break down the door. The pollution inside would overcome your senses and render you inoperative, unconscious. You must remain on deck.

Yet you need to do something to put the ship right, to at least communicate to the person locked in the hold that you are there and able to guide the ship. You keep trying to reach the person inside.

The ship is your body; the person inside is your mind—your conscious mind; the person outside, who can see but is helpless, is your double—your higher or spiritual self, your dream body. The ship, of course, has been built to take its passenger somewhere, but its design renders it difficult to steer from an outside vantage point. The real steering must come from a soul who manages to lock in to it, and, alas, loses his soul in so doing. The double is there to remind the locked-in soul that there is an outside, a direction for the ship, and a purpose for it. However, if, in running the ship, the person inside fails to communicate, fails to receive direction from the dream body outside, then the ship will fail to reach its destination. It will simply flounder and finally sink when it runs aground.

Illness is the same kind of message: It works by rotting the ship's foundation in order to open the door and let in some light, giving the soul needed spiritual breath.

Each human body is troubled by two equal and opposite drives, each trying to take control in order to survive. One is the drive to *do*; the other is the drive to *be*. Doing is action, movement, a steering of life's course; being is passivity, a motionless, timeless, spaceless self—"Don't just do something, stand there."

Passivity is necessary in order to hear the outer soul, which cannot enter the hold without being overcome by the stench of daily life. The activity is necessary in order to carry out the driving force of the universe: creation and evolution. To create inevitably means to contaminate, to pollute, the body, the earth, the universe. Perfection is sterile. To end all creation is to reach the state of absolute timelessness, where and when there is nothing to do. And that is no fun at all.

Our bodies reflect the do-be-do-be-do of the Sinatra melody—the creation and the rest from that creation. Illness, I propose, results from a disharmony produced when parallel universes sing out of tune because they fail to listen to each other.

All of us in Western society have lived with the illusion that we have no bodies. Until illness strikes, we more or less move this bondage of flesh around, ignoring it as much as possible unless we are horny, hungry, or have to go to the toilet.

Tuning in to it seems impossible unless we have had biofeedback or body awareness training—or unless we have had a physical setback which commands us to listen in with all ears.

Instead of looking at illness as something wrong, think of it as something finally going right: The inner soul is finally seeing some light. The hold has rotted through and the outer soul (or double) now is able to communicate. Surprise! The inner and the outer souls have the same face!

CHAPTER 32

A Quantum Model of the Ego, Stress, and Its Relief

In this chapter, I examine what I believe to be *the* crucial element in human illness: the rise of human stress. More than any other causative factor, I believe that our thoughts are the primary causes of stress. More important than any medical care is the concept of prevention through good mental habits. By good mental habits I mean thinking positively about every situation that you encounter. In the previous chapter, this approach amounted to learning to "hear" the healers among your other parallel-world bodies. This means listening to and heeding the aches and pains of your body. This could be particularly true in the so-called universal diseases mentioned earlier.

How can we provide for a more healthful human existence? Human existence depends on human thought, and thought depends on our self-concepts—our egos. I believe that a study of the rise of the human ego is paramount in order to grasp just why and how we make ourselves ill, and often how we age ourselves faster than need be.

The concept of the ego has undergone many revisions. I have reverted to the primary definition as first given by Sigmund Freud and other profound and modern conceptions of it as given by spiritual teachers such as Da Free John, J. Krishnamurti, the disincarnate entity Seth, and Paramahansa Yogananda. Together, these minds have provided me with an insight into the quantum physics of the construction of the ego.

The Ego

We owe the basic concept of the ego to Sigmund Freud, who saw it as a construction within the psyche (or soul). This construction arose out of a previous psychic construct called the *id*. The id is the oldest psychical apparatus; it arose from a basic Freudian assumption that every human being has an inner mental life that is a function of a psychical apparatus. This apparatus has both spatial and temporal extent. Freud never hinted as to where the id existed or from what it was constructed. The id, as Freud stated, "contains everything that is inherited, that is present at birth, that is laid down in the constitution—above all, therefore, the instincts, which originate from the somatic organization and which find a first psychical expression here [in the id] in forms unknown to us."

The ego arises out of the id because the id must interface with the "real" world of stimulations and sensations. That portion of the id called the ego undergoes a special transformation. From the surface of the cortex itself, that is, a cortical layer, a special organization has arisen which acts as the intermediary between the id and the outside world's stimulations. The ego, in consequence of the preestablished connection between sense perceptions and muscular action, has voluntary movement at its command. It has the task of self-preservation, a task that it can perform by becoming aware of stimulation, by memory, by avoidance of stimulation, by adaption, and by learning. It operates within the id by performing the task of gaining control over the demands of the id (the instincts), by choosing which demands to satisfy, by postponing satisfactions of the id, and by consideration of tensions produced by stimuli. It is further able to differentiate between these tensions in terms of what is felt as pain (or unpleasure) and pleasure. The actual sensing of pleasure exists as a vibrational pattern between two poles of tension called the pain and pleasure points. An increase in tension is felt as pain, while a decrease is felt as pleasure.

In his theory of the instincts, Freud decided that the main tensions arose not between the points of pleasure and pain, but between two basic *instincts*: love, and death.

All the above theory we owe to the genius of Freud. In recent times, ego has become a major word in Western vocabulary and a vibrant consideration for the rational human.

The Ego as Seen by Spiritual Teachers

Da Free John considers the ego as a devastating construct that keeps human beings from realizing their God-selves. He points out that we each live our lives in egoic stress; the ego is a process of self-possessed physical, emotional, and mental reaction to the circumstances of life. The ego's action is stress production. Stress, he explains, is easy to trigger through the frustration of self-motion or through the fear of taking that motion. The stress is released either through taking that motion, or by the relaxation or release of the frustration reaction. In order to accomplish this release, we must learn to notice when stress is arising, a major insight gained through self-knowledge. Although noticing when stress arises sounds simple enough, few of us notice when we are becoming stressed. Noticing that you are stressed and feeling that stress at the same time are like the proverbial rubbing your stomach and scratching your head simultaneously. In a typical situation, someone may say something to you that is particularly upsetting; you probably react by getting angry or in some way feeling depressed. Although you are certainly aware of how you feel, you normally are not aware that a stress has arisen as a result of these feelings. In other words, you feel, but you don't know you feel. I have seen (as I am sure you have also), time and time again, people who obviously are very angry, but who answer in the negative when asked whether they are angry. Of course, our first impression is that they are lying to themselves: They must "see" that they are angry. But look again. Most people don't consciously lie to themselves; they only appear to be lying because it seems so obvious to others that they *must* be lying.

I suggest that the knowledge of a feeling and the feeling itself are complementary to each other. The knowledge that you are having a feeling will alter that feeling in an unpredictable manner, certainly changing it. Like the position and the momentum of a particle, you cannot simultaneously feel and know you are feeling at the same time.

In the same way, you cannot be stressed and know you are stressed at one and the same time. Thus, becoming aware of stress alters the stress. An amusing example comes to mind. When making love, and obviously enjoying yourself, you may say to yourself, "Boy, am I having a good time." Just thinking about what you are feeling alters the feelings. (I am sure you know what I mean.)

Paramahansa Yogananda, in *Autobiography of a Yogi*, describes the ego as the root cause of dualism—the seeming separation between man and his creator. *Ahankara* (desire) brings human beings under the sway of *maya* (cosmic delusion), by which the subject (ego) falsely appears as object: The creatures imagine themselves to be the creators.

J. Krishnamurti, in conversation with physicist David Bohm in their book, *The Ending of Time*, talks about senility and the brain cells. They argue that our brains, when looked at collectively, are very old. A human brain is not any particular brain; it doesn't belong to anyone. It, instead, has evolved over millions of years.

Consequently, there are built-in patterns for success and survival that exist today but that may be outmoded. One of these patterns is the ego and its tendencies.

The disincarnate entity Seth, in Jane Roberts's book *The Unknown Reality*, describes the ego as specialized in expansions of space and its manipulations. The ego arose in tribal environments as a necessary specialization; it enabled data from the senses to be differentiated emotionally and otherwise. Tribes formed in which members were considered as being either inside or outside the tribe. This tribal consciousness was the first group ego. Later, consciousness was not able to handle the tribal ego as it was, and individuation began to take place. This process depended on cooperation between the members of the tribe. Thus, the individual ego arose as an agreement between the tribal members.

There is much that can be augmented here. In future writings, I hope to spell out in more detail how quantum physics can explain the human ego. Here I want to offer the first, as far as I can tell, quantum physical model of stress and the rise of the ego. I believe that we can grasp the concept of the ego more firmly if the concept can be described, though only metaphorically for the time being, in scientific terminology.

A Quantum Physical Model of the Ego

To begin with, many physicists believe that all matter is composed of trapped light. This belief is embodied in the famous $E = mc^2$ equation of Einstein. The notion here arises from the fact that every particle of matter has a mirror-image particle of antimatter. When a particle of matter interacts with its mirror-image particle, it undergoes a process called matter/antimatter annihilation, the result of which is the production of light or massless energy. Thus, matter is trapped light.

In my earlier book *Star Wave*, I speculated that human feelings such as love and hate could be described in terms of simpler and more primitive base feelings. These base feelings were found in the matter-light transformations of electrons. Hate, for example, was explained as a quantum statistical property of electrons, connected with the fact that no two electrons will ever exist in the same quantum state. Love was explained in terms of the quantum statistical behavior of light particles—photons: All photons tend to move into the same state if given the chance; thus, in a physical sense, the phrase "light is love" is no exaggeration.

And in a similar sense, the reason we all suffer from loneliness and other human types of pain connected with our material bodies is due to the hate properties of electrons. These electrons are, in a certain sense, trapped light. I speculate that electrons "feel" some form of quantum suffering. This is not like our suffering, but our suffering arises from theirs. This suffering is a desire to become light once again. When an electron meets with its anti-matter partner, the positron, the two particles annihilate each other, producing the extremely high-frequency light known as gamma rays.

I believe that all of our human feelings and emotions are rooted in these simpler physical properties of matter and that human feeling can be explained by looking at the group properties of many electrons in the human body.

Each electron in each atom of matter possesses well-defined classical properties, such as electrical charge, mass, spin, magnetic moment, inertia, and location in space. The last attribute is, however, suspect. This suspicion is due to the wave-particle duality of all matter—the quantum nature of the physical world. Electrons can be imagined best as "events with attributes," rather than as objects with properties. The electron, in other words, is a construct of human thought. Since human thought is limited to immediate sense impressions and since quantum physics is based upon a world beyond such impressions, none of us can know what an electron really is. I take this as a necessary and sufficient condition for the existence of the physical world.

The paradox of wave-particle duality is extremely important in understanding the rise of the ego. First, however, let me talk about the quantum physics of the id.

The id, according to my view, is totally governed by the mathematical framework of quantum physics. In brief, the id is a Hilbert space—a mathematical space of infinite dimensions, each dimension comprising a quantum state of existence or measurable attribute. An ordinary atom of hydrogen, consequently, has an infinite number of quantized energy states. These states are the dimensions of its Hilbert space. Since the id, according to Freud, is composed of timeless states, I postulate that these Freudian states are nothing more or less than energy states of the complex human-energy system. I further offer that energy states and emotional states are one and the same. Thus, anything physical expresses feeling when it undergoes energy transformation.

Not everything expressed energetically, however, is felt. What we call the rise of feeling is a result of transformation of energy states, and this transformation requires a rather complex network of sensations to arise in the first place. This network is seen in the human being as the nervous system and brain.

Perhaps it would be useful to emphasize that I don't mean the same things by the terms "feelings" and "sensations." I refer to the way in which Carl Jung used them. Sensations are easier to speak about; they involve movements of electrons or other electrically charged particles from one place to another. A sensation implies the existence of a disturbing event or factor, such as the prick of a pin on the skin or a molecule of sugar landing on a taste bud. Sensations include vibration, heat, cold, taste, smell, sight, and sound. For a sensation to arise, some location in the body must register it. The skin, consequently, registers the location of pin pricks or heat; the tongue registers the sensation of taste. Sensations correspond to the quantum physical operation of locating some particle at some registering device in the body, usually a nerve ending.

Feelings are different from sensations; they are more elusive to describe, and they correspond to the evaluation of sensation. A particular vibration, therefore, may be evaluated as a good "vibe" when it comes from a loving friend or as a bad one when it comes from someone who is angry. Feelings of "wonderfulness" upon tasting good food or the elation of comfort when feeling a warm fire on one's bottom after "skiing" down a snowy hill on that most-delicate part of the anatomy are other examples. The rubbing movements of a partner against our skin, the heat of the fire, are just sensations. And while feelings involve sensations, they are not dependent on them. You can have feelings without sensations causing them. These occur in dreams or in remembrances. Feelings correspond to waves in my quantum model of psychology, whereas sensations correspond to particles.

Feelings would not be felt if the nerve cells did not have membrane boundaries. Felt feelings are due to bodily sensations caused by electrical changes in the boundaries of nerve cells (see Chapter 15). Thus, felt feelings are feelings transformed into sensations. When feelings are felt or expressed, the body will experience sensations. It won't be possible, however, to completely determine those sensations if that feeling is strong. This is similar to certain aspects of the complementarity between particle locations and wave movements associated with the quantum physical description of matter. In a similar manner, particles "seen" in a physics experiment are energy states transformed into positions or locations in space by an electrified, locating/sensing instrument, such as a bubble or spark chamber.

The ego is a construct arising out of energy transformations expressed as bodily sensations. Therefore, the ego does not simply exist in the brain; it exists wherever cells have boundaries and whenever those boundaries are capable of undergoing spatial change. In a sense, every cell has an ego. Any living entity that has a surface, in fact, will have an ego. Of course, animals have egos; and in this same sense, so do plants, amoebas, and other single-celled forms of life.

To grasp how the ego arises and undergoes change or transformation, we need to look again at that most important factor of quantum physics: the observer effect.

THE OBSERVER EFFECT

According to the observer effect, the act of observation is always accompanied by a sudden, irreversible jump in the thing observed. When an atom's light is seen and its energy measured, therefore, the atom, having existed previously in a timeless state of "no energy"—or, in other words, as a superposition of all possible energy states with equal probabilities—suddenly takes on a specific energy state when the light photon is emitted.

This sudden realization is a noncontinuous quantum jump from one state to another in what mathematicians call Hilbert space. Associated with these jumps, or *qwiff-pops* as I called them earlier, is a means, or a mechanism. This is found in the quantum physical postulate of complementarity.

Complementarity says that every measurement obtainable, every thing capable of being observed, is associated with a specific quantum physical wave function, or qwiff. This qwiff, in turn, is composed of the sum of other qwiffs. Each term in the sum is itself a qwiff, representing a complementary observable. The qwiff representing the energy state "feeling good," by way of example, is composed of complementary thought states—states that have no feeling but that themselves are thoughts. These thoughts include "I feel good" and "I feel lousy." Thus, when you begin to question your feelings, which means bringing into power the apparatus for thinking instead of the complementary apparatus for feeling, you have "mixed feelings" about how good you actually feel.

Returning to lovemaking once more, a familiar example comes to mind. While you are making love and your bodily sensations are transformed into feelings, you rarely think about it. And by not thinking you experience the most sublime and beautiful feelings; indeed, that's what lovemaking is all about. But the moment you begin to think to yourself, "I wish my partner would do this," or, "I wish that I felt that," your sensations will continue but your feelings will be completely changed. Another example occurs in listening to someone who possesses obvious charisma giving a talk. If the speaker's manner is "sexy" or "warm," we often find ourselves "swept away" by the words, even though we might react with disdain if we read those same words. In this case our feelings are aroused and our logical ability to follow the content is diminished. Through just such complementary devices dictators have come to power, perhaps even presidents. I am sure that many of the old clichés contain similar examples; a familiar old Yiddish saying has it that "When the penis arises, the brain goes to sleep."

Thinking and feeling are complementary operations associated with a whole spectrum of thoughts and feelings. So are sensing and intuiting. Intuiting depends on bodily sensations and is complementary to them in the same manner that feelings depend on thoughts and are associated with them. That raising of the hair on the back of the neck is the sensational experience of the intuition that someone is behind you or that something is likely to occur in the near future. When the psychical apparatus chosen is "thoughtful," feelings are altered and often undefinable. So too when that apparatus is "feeling," thoughts are altered and often undefinable. Every observation causes a repression that is associated with feelings when thoughts arise and, conversely, associated with thoughts when feelings arise.

Thus, associated with every observation there will be a repression of complementary observations. The apparatus that causes these choices to be made lies within and is constructed from the id. It is the ego, and here is how it arises.

The Rise and Creation of the Ego, Stress, and Pain

The province of the ego is the body-mind. The body-mind is a concept defining the boundary existing between the body and the mind. If we look at this boundary carefully, it begins to blur, and the distinction between what we call the body and what we call the mind begins to vanish. The body-mind is probably a very old construct. Animals possess well-defined body-minds because their thought patterns are probably simpler than ours. The body-mind of an amoeba, for example, is the animal itself. It is only in our own case that the differentiation between body and mind becomes so complex, because of our ability to express thought and other abstractions to one another.

"The unenlightened body-mind," according to Da Free John, "is founded on the actions of *self-contraction* [italics mine]. The self-contraction is expressed as the differentiation of self from the Transcendental Source-Condition and from every other form of presumed not-self, and is likewise expressed via the independent definition of self and the constant concern and search for independent self-preservation. The self-based or self-contracting and self-preserving conception of existence is manifested via the psychology of fear and conflict relative to all that is not-self."

Yogananda describes the various links between normal mental modifications and functions in the *sankhya* and *yoga* systems. According to him:

> The different sensory stimuli to which we react, tactual, visual, gustatory, auditory, and olfactory, are produced by vibratory variations in electrons and protons.

These depend, in turn, on what is called the *maya* of duality. *Maya* means, literally, "cosmic illusion;" it also means "the measurer." Thus, *maya* is the magical power in creation by which limitations and divisions apparently are present in the Immeasurable and the Inseparable.

Both Da Free John and Yogananda point out the separation that arises between something called self and not-self. It is possible to realize this separation in one's own consciousness. During a quiet time, when you have managed to reduce distractions to a minimum, try to become aware of the self that is aware of the common senses you experience. For example, close your eyes for a brief moment. Watch as thoughts arise in your mind, but pay them no heed. Just let them pass through by becoming aware that you are having thoughts. As you continue this process, pay attention to the process of thinking. You will become aware of the division between thinking and the thinker—the self (or thinker) and the not-self (thoughts).

Once each of us identifies with our thoughts, sensations, feelings, or intuitions, we begin to play *maya*'s game with a vengeance. By returning to the timeless, thoughtless, feelingless, and intuitionless state, we cease from this game but nevertheless *live*.

Egocentric existence is thus founded on a fundamental illusion that each and every being will exist forever and separately for all time. This illusion causes all kinds of misfortune involving control of being, existence, and the lives of others. It also causes fear, sorrow, and anger simply because at its heart there lies a fundamental error.

The body quantum model I propose here is quite simple; it is based on the observer effect upon any quantum physically confined particle, group of particles, cell, or neuron. It also applies at the human level.

Consider an electron, otherwise completely free of interaction with any other thing, confined to move within a certain space. In quantum physics, such a system is called "a particle in a box" (see Figure 45). The particle, because it is confined, will not be able to exist outside the volume of the box. Each time it reaches the boundary, speaking in classical physics terms, it bounces off and returns to the confines of the enclosed box.

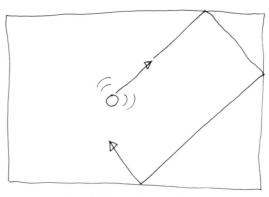

a classical particle in a box

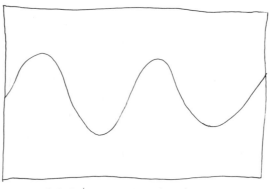

a quantum particle in a box

FIGURE 45.

In the new physics language, the particle does not ever possess a well-defined location inside the box. Such a state of location would provide it with an enormous uncertainty in momentum or energy, resulting in its powerful escape from the box with explosive energy. Instead of dealing with the particle's certain position at each instant of time, we talk about the particle's qwiff, or probability wave, that vanishes at the box's boundaries. This means that the probability of locating the particle at or on the boundary is zero.

The box boundaries act on the particle qwiff in much the same manner that pegs or struts act on a violin string. The pegs hold the string down, keeping the pegged portions of the string from vibrating, which allows for only certain musical notes being playable on the string. The string must vibrate with a certain number of half-wavelengths.

According to quantum physics, only certain energy states for the boxed particle can be experienced. Consequently, because of the boundary conditions, the particle will exist only in certain energy states.

If the boundaries of the box were to be moved farther apart, increasing the box's volume, the energy states would grow closer together. If, on the other hand, the box's boundaries were to move together, decreasing the volume, the energy states would grow explosively apart. In other words, changes in energy, expressed by light being given off by the electron, would be greater the smaller the confines of the box. An electron free to roam a box as big as a room would undergo minuscule energy changes whenever it radiated, but an electron confined to an atom would undergo much larger energy changes.

The ego arises or is constructed from bounded systems in much the same manner as particles in boxes. At first, with primary sense impressions, the ego boundary is quite large; no differentiation is made between the boundaries of the ego and the boundaries of the whole universe. The infant, for example, does not differentiate itself from the universe that he or she is surrounded by: He or she is all-knowing. But then events are imposed upon the sensory apparatus associated with pain or pleasure. The child's ego is now forming.

In my simplified model of this, the original ego state—the ego you had before you were born—exists in a stable, bounded, but quite large space—the whole universe. But the first experiences quickly diminish your ego down to immediate surroundings. In this space, energy levels are close together and transformations between them take place, resulting in sensory experience. But next comes a vital and necessary process: the patterning of knowledge. A separation occurs and an artificial boundary arises between the outside world and the inside world. This separation is completely artificial, and it represents an act of conscious awareness, an act of knowing. This action of knowing, which sounds epistemological, is actually ontological in nature: The world of the child has been separated into a knowing and a known.

The actions of the ego, then, are similar to the actions of the particle in the box (see Figure 46). When a learning experience occurs, the ego contracts in the same way that the boundaries of the box contract. (Imagine an amoeba encountering a foreign object.) If the box contraction is rational—meaning that the ratio of the lengths of the box before and after contraction are fractions made of one integer dividing another—the energy state of the particle in the box will be stable. In the ego's case, we say that the egoic contraction reproduces an experience of the world. It contains a representation of the experience in much the same way that a miniature object reproduces the original object.

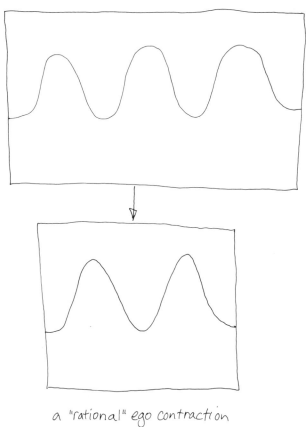

a "rational" ego contraction

FIGURE 46.

If the box contraction is irrational—meaning, the ratio of sides of the box before and after contraction is an irrational number, for example, the square root of 2—then the energy of the particle in the contracted box will not be stable (see Figure 47). The particle will be in a state in which the energy is not determined. The ego will experience a trauma; there will be a memory associated with the state, but it won't be a pleasant one. Furthermore, since in the box case the system has smaller boundaries, the energy states will

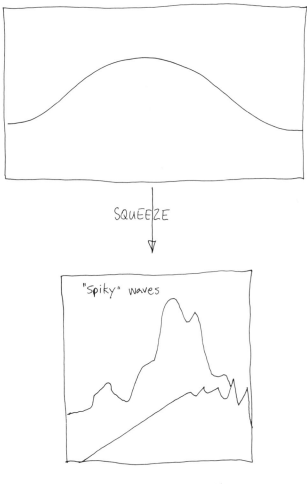

an ego "irrational" contraction

FIGURE 47.

be farther apart, as mentioned above. It then becomes harder to induce changes in the energy. That is why contracted boxes appear more stable. But that means, analogously, that it is difficult to change the ego by energetic means, such as physical activity.

If the ego could expand, it would tend to disrupt the memory pattern, much as an expanded box disrupts the energy states of the box. The ego expansion will break the pattern, making it possible to learn again.

Tension, or the feeling of forces, arise when expansions and contractions take place. In the box example, tension is a physical force on the particle; with the ego analogy, it is the feeling of mounting stress or the release of that stress.

With rational contraction, no spikes of tension occur; indeed, there is nothing to even recall that a change has occurred. Nothing, except that now the particle is confined to a smaller space. Any energy changes it suffers, consequently, will be proportionately larger. Analogously, this results in the contracted ego's undergoing more effort to replicate experience by changes in feeling than before the contraction. In other words, it requires more work and energy to create changes in energy for the boxed particle in the contracted state than in the original state.

It is important to realize that rational contraction arising without tension is probably the way in which learning takes place. The ego has formed without any real awareness that it has taken form. However, the price of no tension is apparent. It now takes more work for the cellular structure to undergo changes in feeling.

Irrational contraction (with tension) creates pain, or spikes of force piercing the boundaries of the cell. Thus, the ego attempts to modify or change that situation either by further contraction or by expansion of the ego. Mind you, I don't mean ego inflation as in the Jungian sense; I mean egoic expansion in the physical sense, that the particle now attempts to occupy a larger space than it had previously.

The Physics of Ego Expansion, Stress Release, and Pleasure

Contraction or differentiation of existence into separate self-possessed units cannot result in anything more than pain or ennui. No pleasure is found in it, and yet self-contraction appears to be a primal drive associated with survival. As we see in my simple model, contraction can lead only to negative emotions associated with pain, such as sorrow, anger, and all kinds of destructive acts.

Yet it is necessary for a memory to be recorded. By making each egoic cell smaller, the world becomes larger and more unfriendly. Each unit, in diminishing itself, is actually programming itself for self-annihilation, much like the main character in "The Incredible Shrinking Man." Yet we humans persist in this illusion. Remember, with ego shrinkage it takes more work to feel anything at all. Shrinking egoic existence down to the size of atoms becomes the desired goal ultimately of egoic existence. It is no accident that we are now physically able to shrink our egos this small at this moment. Any series of nuclear explosions accomplishes the job.

But what can we do about contraction? The answer, according to quantum physics, is expansion. What happens when a particle in a box is offered more space to wander in? Here we see that there is only one effect—the rise of tension that can be viewed as pleasure, pain, and possibly even enlightenment.

Expansion is, however, not the opposite of contraction. No expansion results in tension-free existence. There always will arise forces when the quantum physical system undergoes an expansion of its boundaries. The previously confined timeless wave, upon expansion, finds itself in a larger space, and like a wave, it sloshes back and forth in that space, attempting to reach an equilibrium it never finds. (See Figure 48, page 278.)

I believe that these forces give rise in the ego to physical pleasure simply because they are vibratory tensions or rolling sensations. The frequency of these vibrations of tension, of course, depend on the size of the cell and the mass of the particle that is confined in it.

This difference constitutes the distinction between pleasure and pain. In other words, pleasure is the change in time of tensions or forces arising within cellular units or egoic structures in a more or less continuous or smooth manner. The closer the pattern resembles a sinusoidal wave of intensity, the greater the pleasure. On the other hand, spikes of tension, which always occur under irrational contraction, cause too much change in the cell over too short a time period, the result being the wincing of pain or anger.

The distinction between pleasure and pain is simply one of frequency. With a change in frequency, the ego senses a threshold somewhere between pleasure and pain. This feeling might be the result of an explosive initial state of high energy causing the sudden feeling of enlightenment—a sense somewhere between pain and pleasure.

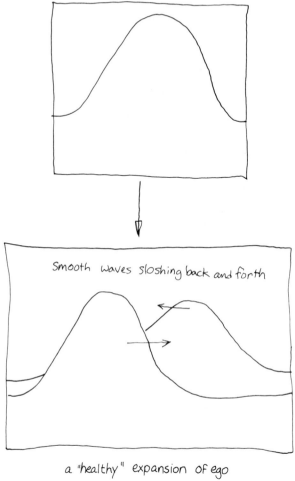

a "healthy" expansion of ego

FIGURE 48.

The Body Quantum's Ego

In this chapter, we have seen how quantum physics can be extrapolated from the world of dead inert matter into the world of the living body-mind. Here I propose that the ego is a quantum physical construction of our brains and nervous systems similar to the quantum world of a particle in a box. According to Freud, our brains possess psychical apparatuses, but he does not hint at what these apparatuses might be. Specifically, he doesn't tell us how our id and our ego are constructed. In my view, the id is a mathematical space, called a Hilbert space, composed of the many possibly different energy states of the physical nervous system. The ego arises out of the id as a physical operation, tending to choose which states are observed. The ego, then, is a surface in the abstract space; it is also a surface in the body. The way the ego operates is through contraction and expansion of the physical neurons' surfaces. Rational contraction will reproduce physical experience—memories; irrational contraction will result in painful recall. Ego expansion, which means physical expansion of the neural surfaces, results in pleasure. Thus, the ego is limited by pain and pleasure and exists between expansion and contraction.

Relief of stress always is associated with pleasure and, therefore, with an expansion of neural boundaries. A similar expansion takes place in all the cells of the body when stress is relieved. Because there is complementarity between thoughts and feelings, the ego will not ever be successful in reproducing all the possible physical experiences of the world. Human unhappiness results from a failure to grasp this fact.

CHAPTER 33

The Transformation of the Whole Body-Mind

In the previous chapter, I tried to explain by a simple model how the ego arises and how that rise is associated with self-contraction. By expansion the energy changes of a particle in a box are actually smoother. This means that transitions between energy levels become more continuous as the boundaries expand and, conversely, that energy transitions become more discontinuous as the boundary diminishes.

A Model for Health

Energy transitions in the box are associated with changes in feelings in the ego. It follows, then, that the smaller the egoic structures, the smaller the egos are literally, the more painful will be the learning experience associated with replication of experience or memory. Conversely, the larger the ego structure associated with new experiences and feeling, the easier the learning of new experiences and the gentler the changes in feelings.

The key insight here is the value of wholeness. By expanding the ego to include the whole body, thus making the whole body the body-mind, learning and pleasurable existence are manifested. This may seem easy enough, but it is not the usual way we humans learn to operate. Our Western habit of rational analysis tends to break down experience into smaller and smaller units. In modern brain research, we look for units associated with the human senses. We draw the picture of the homunculus—a little human inside us whose sensory apparatus stretches along our cortex, with distorted features corresponding to our senses—proportional to areas of the cortex. We associate learning with areas of the brain.

Yet the enlightened condition transcends the specialization of experience within well-defined boundaries. The smaller we compose our egoic structures by irrational self-contraction, the greater is our discomfort; the more intense our pain, the harder it is to learn and to replicate the outside world; the greater the appearance of the outside world, the less enlightened we become. The enlightened condition transcends egoic existence by expansion of the ego to include or transmit across boundaries erected previously by the actions of consciousness.

These actions can arise as egoic contractions that trap consciousness into smaller egos. Or, they can arise as expansions, momentarily freeing consciousness by making larger units through the fusion of units into each other. I don't mean the simple forming of a collective of egos; I mean the dissolution of egoic boundaries. Whereas before fusion there were two egos in volumes of one cell each, there is now one ego in a volume of two cells. Indeed, they are no longer aware of themselves as separate cells. In a real sense, the ego expansion has made them smarter as one than they were as two.

As ever more cellular units join in a fusion of consciousness and dissolve their self-interest boundaries, the qwiff representing them becomes ever more universal, able to reach across wider and wider regions of space. In this sense, the common form of awareness that we all call consciousness emerges.

Thus, health becomes a simple procedure. To make something healthy, include it; to make something sick, exclude it. Exclusion is going unconscious about that which disturbs you, denying it, ignoring it. Inclusion is paying attention to it. Health is nothing more than the conscious awareness of that which pains us. Pain always arises in exclusion through consciousness forming into smaller units; pleasure, or enlightenment, always arises when "pain-full" units are included in the whole.

A Global Model for Health

If the cell's pain is intense enough, it will exist in a high-energy state. Its expansion will not necessarily be a pleasant experience. The greater the initial energy or feeling of the "unloved" cell, the more intense the experience associated with release from the confines of loneliness.

A similar idea can be applied to human culture. Therein lies the fear of loving relationships—fear to expand human egos beyond our own skins.

According to many spiritual teachers, our society exists in a time of technological sophistication in which the egoic model of humankind and human society is the universal basis for the mind.

281

Gross materialism presents human beings with no options but those of trapped, threatened animals. Thus, as Da Free John puts it, "We each feel fiery, full of desire, frustration, fear, despair, and aggressive reactivity."

Our advertising uses sex as a sales device simply because we are preoccupied with sex. This preoccupation occurs because of the continual rise of the fear of death. No wonder we are so sexy and prone to problems associated with sex, the body, and health. Yet the cure to our problem will not be found in a better diet or a sexier waistline; it lies in our realizing that we continue to promote grand illusion. By always focusing on the material aspects of human survival, we continually are creating an egoic society—a society of materially brainwashed egos. "Security is found in real estate" is the motto of our time, it seems.

But the real estate is not "real." The true security lies instead in "reality estate." That reality is a living universal life current, or qwiffness. It is the "body of God" we refer to spiritually, without any recognition of what that could be. To experience the expansion of ego that I am talking about, we must face the fear of expansion. That fear is experienced as our own deaths. Yet each of us has experienced death whenever we have made a sacrifice. The proverbial mother telling her son, "Tell the truth, George, did you cut down the cherry tree?" caused George's ego to undergo death when he told the truth. On the other hand, lying about it would also have caused George to suffer; by lying, his ego structure would have collapsed even more.

Our society lives by its lies. When we can exist without "continually gorging ourselves with our own creativity because if we don't we shall perish," we will find that stressful existence can literally vanish. A spirit of joining will create transformations of body-minds popping off like popcorn around the planet. With enlightenment, all human suffering will be just so much fodder for Hollywood plots, and instead of seeing humans suffering for real, we can just go to the movies.

Afterword: The Body Future

As I look back over this book, I am again reminded of how magnificent the body human is. In spite of its many shortcomings, it is a remarkable instrument, able to leap short distances in a single bound and to contemplate a universe billions of light-years in diameter.

The ability to move our bodies more perfectly than any machine we have designed so far, while fully conscious of that movement, wherever it takes us, is a miracle. Thinking of ourselves as machines for a moment, reflect on the fact that we are the only machines that enjoy the refueling process! Eating is both a necessity and a great pleasure for those of us able to reap the benefits of our society. This body machine not only enjoys fueling itself, it also is capable of repairing itself. It can take the most complex forms of molecular life, such as are found in plants and animals, and rearrange them into new life forms. Not only does it rearrange them, it causes them to take part in the great adventure called evolution. And as we evolve, so do our protein molecules—our DNA and RNA.

And this machine we live in is extremely sensitive to sounds (an eardrum vibration of one atomic diameter can be sensed), pressures, touches, smells (just a few molecules, perhaps five, and our noses tell us that the gas is on), taste, and sights (three photons will turn you on to a vision).

Our bodies are not only sensitive but our minds have devised the means to detect organs and cells inside us. The body has become transparent, thanks to the inventiveness of physics and technology. It is amazing to think that atomic nuclei in our water molecules are able to send messages to complex computerized sensors telling about the function and the health of areas completely inaccessible to our naked eyes.

And this machine is conscious—that consciousness especially known to us as we take each breath of our lives. Through breathing we become aware of our control of the living process; with each breath come molecules of oxygen to provide the resting ground for tired electrons that have made their way through the body, giving up energy along the way.

But in spite of all that we have come to know about the body, we are still very much impressed with its mysteries, yes, and its magic. Will it ever be possible to really know how humans think or feel or, for that matter, whether life is indeed a miracle? Or will science take away all the fun and magic, leaving us with cold formulas as we take in our last breath of air and say goodbye to this world?

I don't think that will be the case because I am, after all, a quantum physicist and I already see that the body's mysteries are little different from the mysteries of the quantum itself. Perhaps the notion of the *Body Quantum* is more than just a catch title; after all, this body, being part of all bodies throughout time, invented the quantum.

No, in spite of my training as a physicist, I am still naïve and dreamy enough to think that science will not dispel the mystery of life, but, instead, will add new dimensions to that mystery. After reading this book and considering my speculations about consciousness and, in particular, about the notion of quantum consciousness—that quantum physics is the physics ultimately governing all the body's processes, including all of its intelligence—I hope that you, the reader, will sense that mystery in an even deeper way.

We have learned a great deal about the body since the days when Aristotle claimed that the dividing line between the living and the lifeless was invisible. We have gone beyond that naïve picture and discovered a sophisticated division between the living and the lifeless—a division born during the Age of Reason that grew to mighty proportions during our own time, only to be diminished by the lowly quantum. The quantum has blurred the line again. We are made of body quanta. We are the miracle of the lifeless turning into life and the living into the lifeless.

Or, perhaps all things are really alive in the first place. Perhaps as our modern ideas about medicine advance, we will see that Aristotle was right after all. There is no division.

Finally, consider the main role of the tiny but powerful body quantum—the role of bringing to mind the forms and contents of the physical world. To understand how we are conscious of so much going on inside and outside us, yet at the same time of so little (we can't feel our brains or our livers, for example), we must grasp that there exists a new effect in the physical world: The world is capable of observing itself and altering its structure accordingly. Here, in our own bodies, is where the real magic of quantum physics can be felt, thought, sensed, and appreciated. Indeed, the experiencing of any of those beautiful events we call life is due to consciousness and the body quantum's ability to observe itself collectively. We can begin, in this way, to observe beyond our present senses.

Where will this take us? If even some of my speculations are proved right, we are in store for a whale of a trip—provided we keep the planet together. I foresee medical advances involving consciousness not only of organs and cells but even molecules, electrons, and, perhaps, the ultimate—photons of light. (Our retinas already are able to sense individual photons, but our brains are not yet aware of or sensitive to them.) By consciously manipulating whether a particle, such as a protein molecule in a neural membrane, is a wave or not, I expect that we will be able to change our bodies at will. I expect that with that gain in sensitivity and consciousness new messages will be received and our evolution will be speeded up so fast that it will make our heads spin. Perhaps we will be able to heal ourselves purely by thinking positively about ourselves. Perhaps we will be able to regenerate new limbs, increase our intelligence, and even live for 500 years or more.

If we can learn to live together as a species, we will not just survive this world, we will create it as well as other worlds beyond our present dreams. The intelligence of the body quantum is absolutely unlimited.

Appendix: An Activity Basal Metabolism Program for Weight Control

The following computer program is written in a language called BASIC. It will enable you to obtain an activity/passivity profile for your own caloric intake. The value of your BMR (B in the program) per pound of body weight depends on several factors, and may not actually correspond to your own basal metabolic rate. I have included an age factor, roughly corresponding to the table below. Thus, a person 30 years old will have a higher BRM than a person who is 50. The table below indicates an approximate BMR according to age.

BMR Versus Age

AGE	BMR
0	22
0.5	26
1	25
2	23
4	21
8	18
12	16
18	13
70	10

If you would like a copy of the latest updated version of this program on a floppy 5¼" disk suitable for running in an IBM-PC or compatible computer, send your name, address, and a check for $15.00* to:

Youniverse Seminars, Inc.
Suite K-166
7777 Fay Ave.
La Jolla, CA 92037

Allow at least four weeks for delivery.

*California residents, please include 90 cents to cover sales tax.
Note: You will need a copy of BASIC software to run the program. If you don't already have BASIC, please check with your computer software supplier.

```
10  'PROGRAM WEIGHT GAIN AND LOSS USING FAT AS THE BASIS
20  'CONSTANTS
30  'B IS THE BMR IN CALORIES PER POUND PER DAY.
40  'F IS THE CALORIC ENERGY CONTENT OF FAT IN CALORIES PER
    POUND.
50  F=4227 'This is the fat Caloric equivalent in Calories per
    pound of fat.
60  INPUT ''My present weight is'' ;WO
70  INPUT ''My present age is'' ;X
80  'The following calculation corrects the BMR according to
    your age.
90  IF X<5 THEN 120 ELSE 100
100 IF X<=18 THEN B=-X/2+22 ELSE B=-3*X/52+365/26
110 GOTO 130
120 IF X>=2 THEN B=-7*X/6+76/3 ELSE B=25
130 INPUT ''My Caloric intake per day is'' ;IC
140 DW=WO-IC/B 'This is the decrement in weight based on
    start weight;
150 '       and Caloric intake.
160 INPUT ''The number of days I want to consider is'' ;T
170 W=IC/B+DW*EXP(-B*T/F) 'Formula for weight gain or loss
    per day
180 PRINT ''My BMR is'' ;B;'' and my age is'' ;X
190 PRINT ''My passive target weight in this diet is'' ;IC/B
200 PRINT ''My weight will be'' ;W; ''after'' ;T; ''days of
    dieting.''
210 INPUT ''Do you want to include an activity list'';
220 IF A$=''Y'' THEN 240
230 END
240 PRINT ''The activity list is''
250 INPUT ''How many hours per day SLEEPING''; NSLP
260 INPUT ''How many hours per day SITTING PASSIVELY AS IN
    WATCHING TV''; NTV
270 INPUT ''How many hours per day SITTING ACTIVELY AS IN DESK
    WORK''; NDSK
280 INPUT ''How many hours per day STANDING RELAXED''; NSTD
290 INPUT ''How many hours per week WALKING AT 3 MPH
    LEISURELY''; NWLK
300 NWLK=NWLK/7
310 INPUT ''How many hours per week BICYCLING AT 10 MPH'';
    NB10
320 NB10=NB10/7
330 INPUT ''How many hours per week RUNNING AT 5.7 MPH''; NRN1
340 NRN1=NRN1/7
350 A1=IC/F
360 A2=B+NSLP/24*(10.34-B)+NTV/24*(14.69-B)
370 A2=A2+NDSK/24*(26-B)+NSTD/24*(15.55-B)
380 A2=A2+NWLK/24*(32.69-B)
390 A2=A2+NB10/24*(49-B)
400 A2=A2+NRN1/24*(91.7-B)
410 A2=A2/F 'The constant A2 is the active metabolic rate in
    fat pounds;
420 '      'i.e., the metabolic rate divided by F.
430 WA=A1/A2+(WO-A1/A2)*EXP(-A2*T) 'Formula for weight
    after T days.
440 '      'The active target weight is the result of the higher
450 '      'metabolic burn coupled with the Caloric intake. It
460 '      ' is less than the passive target weight.
470 PRINT ''My active metabolic rate, or AMR, is''; A2*F
480 PRINT ''My active target weight in this diet is''; A1/A2
490 PRINT ''My weight will be''; WA; ''after''; T; ''days.''
```

```
500 '_____
510 'If you have a disk drive you can use the remaining part of
520 'This program to save a record of your profile on disk.
530 'If you don't have a disk drive and you do have a line
    printer
540 'Delete 600 through 640 and change PRINT#1, to LPRINT to
    print out
550 'If you don't have a printer ignore lines 580 through 850
560 'by answering N to the question on line 580.
570 '_____
580 INPUT ''Do you want to save a record of this''; B$
590 IF B$=''Y'' THEN 600 ELSE END
600 INPUT ''Name of file''; T$
610 OPEN ''O'', #1,T$
620 PRINT#1, 6500
630 PRINT#1, 70
640 PRINT#1, 1
650 PRINT#1,
660 PRINT#1, ''An activity/passivity weight profile for'';
    T$
670 PRINT#1, ''My BMR is''; B; ''and my age is''; X
680 PRINT#1, ''My present weight is''; WO
690 PRINT#1, ''My Caloric intake per day is''; IC
700 PRINT#1, ''The number of days I want to consider is''; T
710 PRINT#1, ''My passive target weight in this diet is'';
    IC/B
720 PRINT#1, ''My weight will be''; W; ''after''; T; ''days
    of passive dieting.''
730 PRINT#1,
740 PRINT#1, ''My activity list follows''
750 PRINT#1, ''How many hours per day SLEEPING''; NSLP
760 PRINT#1, ''How many hours per day SITTING PASSIVELY AS IN
    WATCHING TV''; NTV
770 PRINT#1, ''How many hours per day SITTING ACTIVELY AS IN
    DESK WORK''; NDSK
780 PRINT#1, ''How many hours per day STANDING RELAXED'';
    NSTD
790 PRINT#1, ''How many hours per week WALKING AT 3 MPH
    LEISURELY''; NWLK*7
800 PRINT#1, ''How many hours per week BICYCLING AT 10 MPH'';
    NB10*7
810 PRINT#1, ''How many hours per week RUNNING AT 5.7 MPH'';
    NRN1*7
820 PRINT#1, ''My active metabolic rate, or AMR, is''; A2*F
830 PRINT#1, ''My active target weight in this diet is''; A1/
    A2
840 PRINT#1, ''My weight will be''; WA; ''after''; T; ''days
    of active dieting.''
850 CLOSE #1
```

BIBLIOGRAPHY

Albert, David Z. "How to Take a Photograph of Another Everett World." Paper presented at the conference New Techniques and Ideas in Quantum Measurement Theory, sponsored by the New York Academy of Sciences, January 21–24, 1986.

Angel, Jack E. *Physicians' Desk Reference*. Medical Economics, 1984.

Bass, L. "A Quantum Mechanical Mind-Body Interaction." *Foundations of Physics* 5, no. 1 (1975): 159.

————. "Biological Replication by Quantum Mechanical Interactions." *Foundations of Physics* 7, nos. 3 & 4 (April 1977): 221.

————. The Mind of Wigner's Friend." *Hermathena: A Dublin University Review* 7, no. 62 (1971).

Becker, Robert O., M.D., and Gary Selden. *The Body Electric. Electromagnetism and the Foundation of Life*. New York: Morrow, 1985.

Bevan, James, Dr. *The Simon and Schuster Handbook of Anatomy and Physiology*. New York: Simon & Schuster, 1978.

Bialek, William, and Allan Schweitzer. "Quantum Noise and the Threshold of Hearing." *Physical Review Letters* 54, no. 7 (February 1985): 725.

Blounston, Gary. "Cancer: The New Synthesis. Prevention." *Science 84* 5, no. 7 (September 1984): 36.

Blum, Harold F. *Time's Arrow and Evolution*. 3d ed. Princeton: Princeton University Press, 1968.

Bodanis, David. *The Body Book*. Boston: Little, Brown, 1984.

Brancazio, Peter J. *Sport Science: Physical Laws and Optimum Performance*. New York: Simon & Schuster, 1984.

Bush, Hayden. "Cancer: The New Synthesis. Cure." *Science 84* 5, no. 7 (September 1984): 34.

California Driver's Handbook. Sacramento, Calif.: Department of Motor Vehicles, 1981.

Cameron, John R., and James G. Skofronick. *Medical Physics*. New York: John Wiley & Sons, 1978.

Chabre, Marc. "From the Photon to the Neuronal Signal." *Europhysics News* 16, no. 5 (May 1985).

Cohen, John, and John H. Clark. *Medicine, Mind, and Man: An Introduction to Psychology for Students of Medicine & Allied Professions*. San Francisco: W. H. Freeman, 1979.

Colgan, Michael. *Your Personal Vitamin Profile*. New York: Quill, 1982.

Collier, R. John, and Donald A. Kaplan. "Immunotoxins." *Scientific American* 251, no. 1 (July 1984): 56.

Combs, C. Murphy. *Webster's Illustrated Family Medical Encyclopedia*. New York: Bonanza Books, 1976.

Comfort, Alex. *Reality & Empathy: Physics, Mind, and Science in the 21st Century*. Albany, N.Y.: State University of New York Press, 1984.

Da Free John. *The Transcendence of Ego and Egoic Society*. Clearlake, Calif.: Johannine Daist Communion, 1982.

————. *A Call for the Radical Reformation of Christianity*. Clearlake, Calif.: Johannine Daist Communion, 1982.

————. *The Transmission of Doubt*. Clearlake, Calif.: Dawn Horse Press, 1983.

————. *Easy Death*. Clearlake, Calif.: Dawn Horse Press, 1983.

————. *Scientific Proof of the Existence of God Will Soon Be Announced by the White House!* Clearlake, Calif.: Dawn Horse Press, 1980.

Deakin, Michael A. B. "The Physics and Physiology of Insect Flight." *The American Journal of Physics* 38, no. 8 (August 1970): 1003.

Dickerson, Richard E. "The DNA Helix and How It Is Read." *Scientific American* 249, no. 6 (December 1983): 94.

Dossey, Larry, M.D. *Space, Time, & Medicine*. Boulder, Colo.: Shambhala, 1982.

————. *Beyond Illness: Discovering the Experience of Health*. Boulder, Colo.: Shambhala, 1984.

Durdin-Smith, Jo, and Diane DeSimone. *Sex and the Brain*. New York: Arbor House, 1983.

Eccles, Sir John, and Daniel N. Robinson. *The Wonder of Being Human*. New York: Free Press, 1984.

Einstein, Xavier. *Trivia Mania: Science and Nature*. New York: Zebra Books, 1984.

Fadiman, James, and Robert Frager. *Personality and Personal Growth*. New York: Harper & Row, 1976.

Feynman, Richard P., Robert B. Leighton, and Matthew Sands. *The Feynman Lectures on Physics: Quantum Mechanics*. Reading, Mass.: Addison-Wesley, 1965.

Feynman, Richard P., Robert B. Leighton, and Matthew Sands. *The Feynman Lectures on Physics: The Electromagnetic Field*, vol. 2. Reading, Mass.: Addison-Wesley, 1965.

Fjermedal, Grant. *Magic Bullets*. New York: Macmillan, 1984.

Fries, James F., and Lawrence M. Crapo. *Vitality and Aging: Implications of the Rectangular Curve*. San Francisco: W. H. Freeman, 1981.

Fullerton, Gary D. "Basic Concepts for Nuclear Magnetic Resonance Imaging." *Magnetic Resonance Imaging* 1 (1982): 39–55.

Gabler, Raymond. *Electrical Interactions in Molecular Biophysics*. New York: Academic Press, 1978.

Gray, Henry. *Anatomy, Descriptive and Surgical*. 15th ed. New York: Bounty Books, 1977.

Haldane, J. B. S. "On Being the Right Size." *Possible Worlds*. New York: Harper & Bros., 1928. See also reprint, 1956, and *The World of Mathematics*, vol. 2. Ed. by James R. Newman. New York: Simon & Schuster, 1956.

Hay, Louise L. *Heal Your Body*. 3d ed. Los Angeles: Louise L. Hay, 1981.

Hayflick, L., and P. S. Moorhead. "The Serial Cultivation of Human Diploid Cell Strains." *Experimental Research* 25 (1961): 585.

Ho, Chien, ed. *Hemoglobin and Oxygen Binding*. New York: Elsevier Bio-medical, 1982.

Hobbie, Russell K. *Intermediate Physics For Medicine and Biology*. New York: John Wiley & Sons, 1978.

Hochstim, Adolf R. "Nonlinear Mathematical Models for the Origin of Asymmetry in Biological Molecules." *Origins of Life* 6 (1975): 317–66.

Hoyle, Fred. *The Intelligent Universe*. New York: Holt, Rinehart & Winston, 1984.

Hubel, David H., and Torsten N. Wiesel. "Brain Mechanisms of Vision." *The Brain: A Scientific American Book*. San Francisco: W. H. Freeman, 1979.

Huisman, T. H. J., and W. A. Schroeder. *New Aspects of the Structure, Function, and Synthesis of Hemoglobins*. Cleveland, Oh.: CRC Press, 1971.

Hunter, Tony. "The Proteins of Oncogenes." *Scientific American* 251, no. 2 (August 1984): 70.

Isaacs, James P., and John C. Lamb. *Complementarity in Biology: Quantization of Molecular Motion*. Baltimore: Johns Hopkins Press, 1969.

Krishnamurti, J., and David Bohm. *The Ending of Time*. San Francisco: Harper & Row, 1985.

Kunz, Jeffrey R. M., ed. *The American Medical Association Family Medical Guide*. New York: Random House, 1982.

Leach, C. S. "The Endocrine and Metabolic Response to Space Flight." *Medicine and Science in Sports and Exercise* 15 (1983): 432.

Lehmann, H., and R. G. Huntsman. *Man's Haemoglobins*. Amsterdam: North-Holland, 1974.

Lehninger, Albert L. *Bioenergetics: The Molecular Basis of Biological Energy Transformations*. New York: W. A. Benjamin, 1971.

Levine, Michael W., and Jeremy M. Shefner. *Fundamentals of Sensation and Perception*. New York: Addison-Wesley, 1981.

Littler, T. S. *The Physics of the Ear*. New York: Macmillan, 1965.

Löwdin, Per-Olov. "Effect of Proton Tunneling in DNA on Genetic Information and Problems of Mutations, Aging, and Tumors." *Quantum Aspects of Polypeptides and Polynucleotides*. Ed. by M. Weissbluth. New York: Interscience, 1964.

————. "Some Aspects of Quantum Biology." *Quantum Aspects of Polypeptides and Polynucleotides*, Ed. by M. Weissbluth. New York: Interscience, 1964.

McKeon, Richard. *Introduction to Aristotle*. 2d ed. Chicago: University of Chicago Press, 1973.

Mercer, E. H. *The Foundations of Biological Theory*. New York: John Wiley & Sons, 1981.

Mindell, Arnold. *Dreambody: The Body's Role in Revealing the Self*. Santa Monica, Calif.: Sigo Press, 1982.

Mindell, Earl. *Vitamin Bible*. New York: Warner Books, 1979.

Morehouse, Laurence E., and Leonard Gross. *Total Fitness in 30 Minutes a Week*. New York: Simon & Schuster, 1975.

Moskowitz, Mark A., M.D., and Michael E. Osband, M.D. *The Complete*

Book of Medical Tests. New York: W. W. Norton, 1984.

Mourant, Arthur E. "Why Are There Blood Groups?" In *The Encyclopaedia of Ignorance: Everything You Ever Wanted To Know about the Unknown.* Ed. by Ronald Duncan and Miranda Weston-Smith. Elmsford, N.Y.: Pergamon Press, 1977.

Nomura, Masayasu. "The Control of Ribosome Synthesis." *Scientific American* 250, no. 1 (January 1984): 102.

Oberg, Alcestis, and Daniel Woodard. "Anti-Matter Probes." *Science Digest* (April 1982): 54.

Pearson, Durk, and Sandy Shaw. *The Life Extension Companion.* New York: Warner Books, 1984.

———. *Life Extension: A Practical Scientific Approach.* New York: Warner Books, 1982.

Perutz, M. F., and H. Lehmann. "Molecular Pathology of Human Haemoglobin." *Nature* 219 (August 1968): 902.

Pritchard, Roy M. "Stabilized Images on the Retina." *Scientific American* (June 1961).

Pullman, Bernard. "Aspects of the Electronic Structure of the Nucleic Acids in Relation to the Theories of Mutagenesis and Carcinogenesis." *Quantum Aspects of Polypeptides and Polynucleotides.* Ed. by M. Weissbluth. New York: Interscience, 1964.

Rand McNally. *The Rand McNally Atlas of the Body and Mind.* New York: Rand McNally, 1976.

Rensberger, Boyce. "Cancer: The New Synthesis. Cause." *Science 84* 5, no. 7 (September 1984): 28.

Roberts, Jane. *The Unknown Reality,* vol. 1. Englewood Cliffs, N.J.: Prentice-Hall, 1977.

Robson, John. *Basic Tables in Physics.* New York: McGraw-Hill, 1967.

Rosenblatt, Allen D., and James T. Thickstun. "Modern Psychoanalytic Concepts in a General Psychology. Part One: Concepts and Principles; Part Two: Motivation." *Psychological Issues* 11, nos. 2 & 3. New York: International Universities Press, 1977.

Ross, John. "The Resources of Binocular Perception." *Scientific American* 234 (March 1976): 80.

Ryan, Regina Sara, and John W. Travis, M.D. *The Wellness Workbook.* Berkeley, Calif.: Ten Speed Press, 1981.

Scientific American Books. *The Brain.* San Francisco: W. H. Freeman, 1979.

———. *The Molecular Basis of Life.* San Francisco: W. H. Freeman, 1968.

———. *The Chemical Basis of Life.* San Francisco: W. H. Freeman, 1973.

Teyler, Timothy J. *A Primer of Psychobiology: Brain and Behavior.* New York: W. H. Freeman, 1984.

Tibbetts, Paul, ed. *Perception: Selected Readings in Science and Phenomenology.* New York: Quadrangle, 1969.

Thomas, Lewis. *The Lives of a Cell: Notes of a Biology Watcher.* New York: Viking, 1974.

Unwin, Nigel, and Richard Henderson. "The Structure of Proteins in

Biological Membranes." *Scientific American* 250, no. 2 (February 1984): 78.

Vannini, Vanio, and Giuliano Pogliani. *The Color Atlas of Human Anatomy.* Trans. and rev. by Richard T. Jolly. New York: Harmony Books, 1980.

Watson, James D., John Tooze, and David T. Kurtz. *Recombinant DNA: A Short Course.* New York: Scientific American Books, 1983.

Webb, S. J. *Nutrition, Time, and Motion in Metabolism and Genetics.* Springfield, Ill.: Charles C. Thomas, 1976.

Weinberg, Robert A. "A Molecular Basis of Cancer." *Scientific American* 249, no. 5 (November 1983): 126.

Wilmore, Jack H., and Dorothy Schefer. "Ideal Weight: New Thinking on Losing, Gaining, Maintaining." *Vogue* (December 1984).

Wolf, Fred Alan. *Star Wave: Mind, Consciousness, and Quantum Physics.* New York: Macmillan, 1984.

———. *Taking the Quantum Leap: The New Physics for Nonscientists.* San Francisco: Harper & Row, 1981.

———. "Trans-World I-Ness: Quantum Physics and the Enlightened Condition." *Humor Suddenly Returns: Essays on the Spiritual Teaching of Master Da Free John.* Clearlake, Calif.: Dawn Horse Press, 1984.

———. "The Quantum Physics of Consciousness: Towards a New Psychology." *Integrative Psychiatry* 3, no. 4 (December 1985): 236.

Wronski, T. J. "Alterations in Calcium Homeostasis in Bone During Actual and Simulated Space Flight." *Medicine and Science in Sports and Exercise* 15 (1983): 410.

Yogananda, Paramahansa. *Autobiography of a Yogi.* Los Angeles: Self-Realization Fellowship, 1973.

Yuspa, Stuart H. "Chemical Carcinogenesis Related to the Skin: Parts I and II." *Progress in Dermatology* (December 1981 and March 1982).

Zicree, Marc Scott. *The Twilight Zone Companion.* New York: Bantam, 1982.

Index

Cells, *continued*
  lysomes, 94
  mitochondria in, 92, 94–95
  nucleic acids in, 93
  nucleoli, 95–96
  nucleus of, 92–93
  prokaryotic, 91–92
  protein molecules in wall of, 92
  ribsomes, 93–94
  See also Cell division;
    Chromosomes
Central nervous system (CNS),
    204–205
  autonomic nervous system (ANS)
    and, 204–205
Centrifuges, 19
Centrioles, 95
Chirality, 82
Chlorophyll, 183
Chromophore, 147
Chromosomes, 98–99, 100–101
  genes in, 98–99
  sex traits and, 98
  See also Cell division
Chyme, 48
Circulation system, 181–82
  gas exchanges, 181–82
  single-cell animals and, 181–82
  surface area of skin and, 181–82
  See also Blood chemistry; Blood
    flow; Blood pressure
Clark, Barney, 167
Clostridum welchii, 191
Cochlea, 132, 137–38
  basilar membrane of, 138
Coenzyme A (CoA), 52–53
Collagen, 12
Color blindness, 149, 156–57
Color vision, 144–46, 149–52
  color blindness and, 149, 156–57
  complementarity and, 152
  dichromatic vision, 154–56
  dispersion and, 145
  focal length and, 145
  future for, 157
  lens-focusing ability and, 145–46
  light, 144–45
  monochromatic vision, 152–54

photons and, 144
primary colors and, 150–51
quantum physics view of, 149–52
trichromatic vision, 156–57
visual bands and, 144
wave aspect of light and, 150–51
wavelength and frequency, 144,
    150, 151–52
Communications, 199–202
  anticipation of future and,
    201–202
  asynchronous flight muscle
    response and, 200
  baseball example, 201
  bumblebees and, 199
  defined, 199
  neural communications, 200–201
Compact bones, 12, 14–16
  Young's modulus (Y) and, 14–16
Complementarity, 152, 269–70
  observer effect and, 269–70
Cones (photo receptors), 149–50
  three sensitivities, 150
Consciousness, 1–2, 30–32, 39–40,
    73–74, 257–58
  cellular construction and, 73–74,
    91
  dream body and, 257
  moving and, 1–2
  muscles and, 30–32, 39–44
  parallel-world hypothesis and,
    257–58
  proprioception and, 39–40
  See also Mind-body interaction;
    Mindful activity; Observer
    effect
Coronary arteries, 178
Correlational bonds, 247–48
  breaking, 247–48
  defined, 247
  observation and, 247
Crapo, Lawrence M., 222, 225
Crick, Francis, 87
Cyanosis, 185
Cytoplasms, 95

Da Free John, 263, 264–65, 270, 282
  on ego, 264–65

309